www.EffortlessMath.com

... So Much More Online!

✓ FREE Math lessons

✓ More Math learning books!

✓ Mathematics Worksheets

✓ Online Math Tutors

Need a PDF version of this book?

Visit www.EffortlessMath.com

STAAR Math Exercise Book for Grade 5

Student Workbook and Two Realistic

STAAR Math Tests

By

Reza Nazari & Ava Ross

All inquiries should be addressed to:

info@EffortlessMath.com

www.EffortlessMath.com

ISBN-13: 978-1-970036-26-8

ISBN-10: 1-970036-26-5

Published by: Effortless Math Education

www.EffortlessMath.com

Description

Get ready for the STAAR Math Test with a PERFECT Math Workbook!

STAAR Math Exercise Book for Grade 5, which reflects the 2019 test guidelines and topics, is dedicated to preparing test takers to ace the STAAR Math Test. This STAAR Math workbook's new edition has been updated to replicate questions appearing on the most recent STAAR Math tests. Here is intensive preparation for the STAAR Math test, and a precious learning tool for test takers who need extra practice in math to raise their STAAR math scores. After completing this workbook, you will have solid foundation and adequate practice that is necessary to ace the STAAR Math test. **This workbook is your ticket to score higher on STAAR Math**

The updated version of this hands-on workbook represents extensive exercises, math problems, sample STAAR questions, and quizzes with answers and detailed solutions to help you hone your math skills, overcome your exam anxiety, and boost your confidence -- and do your best to defeat STAAR exam on test day.

Each of math exercises is answered in the book and we have provided explanation of the answers for the two full-length STAAR Math practice tests as well which will help test takers find their weak areas and raise their scores. This is a unique and perfect practice book to beat the STAAR Math Test.

Separate math chapters offer a complete review of the STAAR Math test, including:

- ✓ Arithmetic and Number Operations
- ✓ Algebra and Functions,
- ✓ Geometry and Measurement
- ✓ Data analysis, Statistics, & Probability
- ✓ ... and also includes **two full-length practice tests!**

The surest way to succeed on STAAR Math Test is with intensive practice in every math topic tested--and that's what you will get in **STAAR Math Exercise Book**. Each chapter of this focused format has a comprehensive review created by Test Prep experts that goes into detail to cover all of the content likely to appear on the STAAR Math test. Not only does this all-inclusive workbook offer everything you will ever need to conquer STAAR Math test, it also contains two full-length and realistic STAAR Math tests that reflect the format and question types on the STAAR to help you check your exam-readiness and identify where you need more practice.

Effortless Math Workbook for the STAAR Test contains many exciting and unique features to help you improve your test scores, including:

- ✓ Content 100% aligned with the 2019 STAAR test
- ✓ Written by STAAR Math tutors and test experts
- ✓ Complete coverage of all STAAR Math concepts and topics which you will be tested
- ✓ Over 2,500 additional STAAR math practice questions in both multiple-choice and grid-in formats with answers grouped by topic, so you can focus on your weak areas
- ✓ Abundant Math skill building exercises to help test-takers approach different question types that might be unfamiliar to them
- ✓ Exercises on different STAAR Math topics such as integers, percent, equations, polynomials, exponents and radicals
- ✓ 2 full-length practice tests (featuring new question types) with detailed answers

This STAAR Math Workbook and other Effortless Math Education books are used by thousands of students each year to help them review core content areas, brush-up in math, discover their strengths and weaknesses, and achieve their best scores on the STAAR test.

Do NOT take the STAAR test without reviewing the Math questions in this workbook!

About the Author

Reza Nazari is the author of more than 100 Math learning books including:
– **Math and Critical Thinking Challenges:** For the Middle and High School Student
– **ACT Math in 30 Days.**
– **ASVAB Math Workbook 2018 – 2019**
– **Effortless Math Education Workbooks**
– **and many more Mathematics books …**

Reza is also an experienced Math instructor and a test–prep expert who has been tutoring students since 2008. Reza is the founder of Effortless Math Education, a tutoring company that has helped many students raise their standardized test scores—and attend the colleges of their dreams. Reza provides an individualized custom learning plan and the personalized attention that makes a difference in how students view math.

You can contact Reza via email at:
reza@EffortlessMath.com

Find Reza's professional profile at:
goo.gl/zoC9rJ

Contents

Chapter 1:
Place Vales and
Number Sense

Topics that you'll practice in this chapter:

- ✓ Place Values
- ✓ Compare Numbers
- ✓ Numbers in Numbers
- ✓ Rounding
- ✓ Odd or Even

Place Values

✎ *Write numbers in expanded form.*

1) Thirty–five	$30 + 5$
2) Sixty–seven	___ + ___
3) Forty–two	___ + ___
4) Eighty–nine	___ + ___
5) Ninety–one	___ + ___
6) Twenty–two	___ + ___
7) Thirty–four	___ + ___
8) Fifty–six	___ + ___
9) Ninety–five	___ + ___
10) Seventy–seven	___ + ___
11) Forty–eight	___ + ___

✎ *Circle the correct choice.*

12)	The 2 in 72 is in the	ones place	tens place	hundreds place
13)	The 6 in 65 is in the	ones place	tens place	hundreds place
14)	The 2 in 342 is in the	ones place	tens place	hundreds place
15)	The 5 in 450 is in the	ones place	tens place	hundreds place
16)	The 3 in 321 is in the	ones place	tens place	hundreds place

Comparing and Ordering Numbers

✎ **Use less than, equal to or greater than.**

1) 23 _____ 34

2) 89 _____ 98

3) 45 _____ 25

4) 34 _____ 32

5) 91 _____ 91

6) 57 _____ 55

7) 85 _____ 78

8) 56 _____ 43

9) 34 _____ 34

10) 92 _____ 98

11) 38 _____ 46

12) 67 _____ 58

13) 88 _____ 69

14) 23 _____ 34

✎ **Order each set of integers from least to greatest.**

15) $7, -9, -6, -1, 3$ ___, ___, ___, ___, ___, ___

16) $-4, -11, 5, 12, 9$ ___, ___, ___, ___, ___, ___

17) $18, -12, -19, 21, -20$ ___, ___, ___, ___, ___, ___

18) $-15, -25, 18, -7, 32$ ___, ___, ___, ___, ___, ___

19) $37, -42, 28, -11, 34$ ___, ___, ___, ___, ___, ___

20) $78, 46, -19, 77, -24$ ___, ___, ___, ___, ___, ___

✎ **Order each set of integers from greatest to least.**

21) $11, 16, -9, -12, -4$ ___, ___, ___, ___, ___, ___

22) $23, 31, -14, -20, 39$ ___, ___, ___, ___, ___, ___

23) $45, -21, -18, 55, -5$ ___, ___, ___, ___, ___, ___

24) $68, 81, -14, -10, 94$ ___, ___, ___, ___, ___, ___

25) $-5, 69, -12, -43, 34$ ___, ___, ___, ___, ___, ___

26) $-56, -25, -30, 18, 29$ ___, ___, ___, ___, ___, ___

Write Numbers in Words

✎ *Write each number in words.*

1) 194 _____

2) 311 _____

3) 256 _____

4) 434 _____

5) 809 _____

6) 730 _____

7) 272 _____

8) 266 _____

9) 902 _____

10) 1,418 _____

11) 1,365 _____

12) 3,374 _____

13) 2,486 _____

14) 7,671 _____

15) 6,290 _____

16) 3,147 _____

17) 5,012 _____

Rounding Numbers

✍ **Round each number to the nearest ten.**

1) 24	5) 11	9) 47
2) 98	6) 35	10) 63
3) 41	7) 84	11) 79
4) 26	8) 70	12) 55

✍ **Round each number to the nearest hundred.**

13) 185	17) 222	21) 670
14) 254	18) 311	22) 563
15) 729	19) 287	23) 890
16) 109	20) 927	24) 479

✍ **Round each number to the nearest thousand.**

25) 1,252	31) 31,422
26) 1,950	32) 12,723
27) 5,235	33) 61,670
28) 3,567	34) 71,290
29) 8,027	35) 50,930
30) 52,512	36) 38,568

Odd or Even

✎ *Identify whether each number is even or odd.*

1) 12 _____   7) 34 _____

2) 7 _____   8) 87 _____

3) 33 _____   9) 94 _____

4) 18 _____   10) 14 _____

5) 99 _____   11) 22 _____

6) 55 _____   12) 79 _____

✎ *Circle the even number in each group.*

13) 22, 11, 57, 13, 19, 47

14) 15, 17, 27, 23, 33, 26

15) 19, 35, 24, 57, 65, 49

16) 67, 58, 89, 63, 27, 63

✎ *Circle the odd number in each group.*

17) 12, 14, 22, 64, 53, 98

18) 16, 26, 28, 44, 62, 73

19) 46, 82, 63, 98, 64, 56

20) 27, 92, 58, 36, 38, 72

Answers of Worksheets – Chapter 1

Place Values

1) $30 + 5$
2) $60 + 7$
3) $40 + 2$
4) $80 + 9$
5) $90 + 1$
6) $20 + 2$
7) $30 + 4$
8) $50 + 6$
9) $90 + 5$
10) $70 + 7$
11) $40 + 8$
12) ones place
13) tens place
14) ones place
15) tens place
16) hundreds place

Comparing and Ordering Numbers

1) 23 less than 34
2) 89 less than 98
3) 45 greater than 25
4) 34 greater than 32
5) 91 equals to 91
6) 57 greater than 55
7) 85 greater than 78
8) 56 greater than 43
9) 34 equals to 34
10) 92 less than 98
11) 38 less than 46
12) 67 greater than 58

13) 88 greater than 69
14) 23 less than 34
15) $-9, -6, -1, 3, 7$
16) $-11, -4, 5, 9, 12$
17) $-20, -19, -12, 18, 21$
18) $-25, -15, -7, 18, 32$
19) $-42, -11, 28, 34, 37$
20) $-24, -19, 46, 77, 78$
21) $16, 11, -4, -9, -12$
22) $39, 31, 23, -14, -20$
23) $55, 45, -5, -18, -21$
24) $94, 81, 68, -10, -14$
25) $69, 34, -5, -12, -43$
26) $29, 18, -25, -30, -56$

Write Numbers in Words

1) One hundred ninety-four
2) Three hundred eleven
3) Two hundred fifty-six
4) Four hundred thirty-four
5) Eight hundred nine
6) Seven hundred thirty

7) Two hundred seventy-two

8) Two hundred sixty-six

9) Nine hundred two

10) One thousand, four hundred eighteen

11) One thousand, three hundred sixty-five

12) Three thousand, three hundred seventy-four

13) Two thousand, four hundred eighty-six

14) Seven thousand, six hundred seventy-one

15) Six thousand, two hundred ninety

16) Three thousand, one hundred forty-seven

17) Five thousand, twelve

Rounding Numbers

1) 20	13) 200	25) 1,000
2) 100	14) 300	26) 2,000
3) 40	15) 700	27) 5,000
4) 30	16) 100	28) 4,000
5) 10	17) 200	29) 8,000
6) 40	18) 300	30) 53,000
7) 80	19) 300	31) 31,000
8) 70	20) 900	32) 13,000
9) 50	21) 700	33) 62,000
10) 60	22) 600	34) 71,000
11) 80	23) 900	35) 51,000
12) 60	24) 500	36) 39,000

Odd or Even

1) Even	8) Odd	15) 24
2) Odd	9) Even	16) 58
3) Odd	10) Even	17) 53
4) Even	11) Even	18) 73
5) Odd	12) Odd	19) 63
6) Odd	13) 22	20) 27
7) Even	14) 26	

Chapter 2:
Adding and Subtracting

Topics that you'll practice in this chapter:

✓ Adding Two–Digit Numbers

✓ Subtracting Two–Digit Numbers

✓ Adding Three–Digit Numbers

✓ Adding Hundreds

✓ Adding 4–Digit Numbers

✓ Subtracting 4–Digit Numbers

Adding Two–Digit Numbers

✎ *Find each sum.*

1)
$$\begin{array}{r} 50 \\ + 18 \\ \hline \end{array}$$

2)
$$\begin{array}{r} 32 \\ + 14 \\ \hline \end{array}$$

3)
$$\begin{array}{r} 45 \\ + 16 \\ \hline \end{array}$$

4)
$$\begin{array}{r} 12 \\ + 12 \\ \hline \end{array}$$

5)
$$\begin{array}{r} 43 \\ + 30 \\ \hline \end{array}$$

6)
$$\begin{array}{r} 34 \\ + 15 \\ \hline \end{array}$$

7)
$$\begin{array}{r} 89 \\ + 7 \\ \hline \end{array}$$

8)
$$\begin{array}{r} 63 \\ + 12 \\ \hline \end{array}$$

9)
$$\begin{array}{r} 90 \\ + 10 \\ \hline \end{array}$$

10)
$$\begin{array}{r} 24 \\ + 12 \\ \hline \end{array}$$

11)
$$\begin{array}{r} 42 \\ + 22 \\ \hline \end{array}$$

12)
$$\begin{array}{r} 23 \\ + 18 \\ \hline \end{array}$$

13)
$$\begin{array}{r} 18 \\ + 25 \\ \hline \end{array}$$

14)
$$\begin{array}{r} 37 \\ + 23 \\ \hline \end{array}$$

15)
$$\begin{array}{r} 56 \\ + 35 \\ \hline \end{array}$$

16)
$$\begin{array}{r} 65 \\ + 40 \\ \hline \end{array}$$

17)
$$\begin{array}{r} 77 \\ + 29 \\ \hline \end{array}$$

18)
$$\begin{array}{r} 59 \\ + 26 \\ \hline \end{array}$$

Subtracting Two–Digit Numbers

✎ *Find each difference.*

1)
$$
\begin{array}{r}
32 \\
-15 \\
\hline
\end{array}
$$

2)
$$
\begin{array}{r}
40 \\
-12 \\
\hline
\end{array}
$$

3)
$$
\begin{array}{r}
67 \\
-17 \\
\hline
\end{array}
$$

4)
$$
\begin{array}{r}
18 \\
-10 \\
\hline
\end{array}
$$

5)
$$
\begin{array}{r}
59 \\
-16 \\
\hline
\end{array}
$$

6)
$$
\begin{array}{r}
89 \\
-20 \\
\hline
\end{array}
$$

7)
$$
\begin{array}{r}
78 \\
-21 \\
\hline
\end{array}
$$

8)
$$
\begin{array}{r}
66 \\
-15 \\
\hline
\end{array}
$$

9)
$$
\begin{array}{r}
87 \\
-45 \\
\hline
\end{array}
$$

10)
$$
\begin{array}{r}
56 \\
-19 \\
\hline
\end{array}
$$

11)
$$
\begin{array}{r}
62 \\
-23 \\
\hline
\end{array}
$$

12)
$$
\begin{array}{r}
47 \\
-20 \\
\hline
\end{array}
$$

13)
$$
\begin{array}{r}
78 \\
-29 \\
\hline
\end{array}
$$

14)
$$
\begin{array}{r}
49 \\
-36 \\
\hline
\end{array}
$$

15)
$$
\begin{array}{r}
82 \\
-38 \\
\hline
\end{array}
$$

16)
$$
\begin{array}{r}
97 \\
-45 \\
\hline
\end{array}
$$

17)
$$
\begin{array}{r}
89 \\
-57 \\
\hline
\end{array}
$$

18)
$$
\begin{array}{r}
95 \\
-73 \\
\hline
\end{array}
$$

Adding Three–Digit Numbers

✍ *Find each sum.*

1)
```
  234
+  56
─────
```

2)
```
  523
+ 134
─────
```

3)
```
  345
+ 167
─────
```

4)
```
  460
+ 120
─────
```

5)
```
  432
+ 430
─────
```

6)
```
  235
+ 150
─────
```

7)
```
  789
+  57
─────
```

8)
```
  863
+ 340
─────
```

9)
```
  956
+  89
─────
```

10)
```
  235
+ 112
─────
```

11)
```
  156
+ 117
─────
```

12)
```
  278
+ 190
─────
```

13)
```
  345
+ 125
─────
```

14)
```
  420
+ 120
─────
```

15)
```
  575
+ 234
─────
```

16)
```
  489
+ 354
─────
```

17)
```
  621
+ 213
─────
```

18)
```
  683
+ 293
─────
```

Adding Hundreds

✎ *Add.*

1) $100 + 100 = ---$

2) $100 + 200 = ---$

3) $200 + 200 = ---$

4) $300 + 200 = ---$

5) $100 + 300 = ---$

6) $200 + 400 = ---$

7) $400 + 100 = ---$

8) $500 + 200 = ---$

9) $300 + 500 = ---$

10) $400 + 700 = ---$

11) $400 + 600 = ---$

12) $500 + 400 = ---$

13) $900 + 100 = ---$

14) $100 + 700 = ---$

15) $500 + 100 = ---$

16) $200 + 800 = ---$

17) $800 + 100 = ---$

18) $700 + 100 = ---$

19) $100 + 300 = ---$

20) $500 + 500 = ---$

21) $400 + 400 = ---$

22) $300 + 400 = ---$

23) $500 + 700 = ---$

24) $800 + 600 = ---$

25) If there are 600 balls in a box and Jackson puts 500 more balls inside, how many balls are in the box?

_____ balls

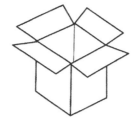

Adding 4–Digit Numbers

✎ *Add.*

1)
```
   1,158
 + 6,687
 ───────
```

2)
```
   5,188
 + 1,298
 ───────
```

3)
```
   5,756
 + 2,712
 ───────
```

4)
```
   3,239
 +2,562
 ───────
```

5)
```
   4,257
 +5,194
 ───────
```

6)
```
   6,215
 +2,189
 ───────
```

7)
```
   3,119
 +1,245
 ───────
```

8)
```
   5,320
 +2,765
 ───────
```

9)
```
   4,890
 +4,567
 ───────
```

✎ *Find the missing numbers.*

10) $1,145 + \_\_ = 1,276$

11) $500 + 1,000 = \_\_$

12) $3,200 + \_\_ = 4,300$

13) $455 + \_\_ = 1,755$

14) $\_\_ + 720 = 1,250$

15) $\_\_ + 670 = 2,230$

16) David sells gems. He finds a diamond in Istanbul and buys it for $3,433. Then, he flies to Cairo and purchases a bigger diamond for the bargain price of $5,922. How much does David spend on the two diamonds?

Subtracting 4–Digit Numbers

✎ *Subtract.*

1) $\begin{array}{r} 2{,}230 \\ -\,1{,}112 \\ \hline \end{array}$

4) $\begin{array}{r} 8{,}519 \\ -\,5{,}422 \\ \hline \end{array}$

7) $\begin{array}{r} 8{,}756 \\ -\,6{,}712 \\ \hline \end{array}$

2) $\begin{array}{r} 3{,}115 \\ -\,1{,}980 \\ \hline \end{array}$

5) $\begin{array}{r} 6{,}222 \\ -\,4{,}331 \\ \hline \end{array}$

8) $\begin{array}{r} 9{,}290 \\ -\,3{,}829 \\ \hline \end{array}$

3) $\begin{array}{r} 4{,}976 \\ -\,2{,}678 \\ \hline \end{array}$

6) $\begin{array}{r} 7{,}821 \\ -\,3{,}212 \\ \hline \end{array}$

9) $\begin{array}{r} 5{,}117 \\ -\,4{,}216 \\ \hline \end{array}$

✎ *Find the missing number.*

10) $2{,}223 - \underline{\quad} = 1{,}120$

13) $2{,}300 - \underline{\quad} = 1{,}250$

11) $3{,}574 - \underline{\quad} = 2{,}245$

14) $3{,}780 - 1{,}890 = \underline{\quad}$

12) $1{,}124 - 578 = \underline{\quad}$

15) $2{,}880 - 2{,}560 = \underline{\quad}$

16) Jackson had $3,963 invested in the stock market until he lost $2,171 on those investments. How much money does he have in the stock market now?

Answers of Worksheets – Chapter 2

Adding two–digit numbers

1) 68	7) 96	13) 43
2) 46	8) 75	14) 60
3) 61	9) 100	15) 91
4) 24	10) 36	16) 105
5) 73	11) 64	17) 106
6) 49	12) 41	18) 85

Subtracting two–digit numbers

1) 17	7) 57	13) 49
2) 28	8) 51	14) 13
3) 50	9) 42	15) 44
4) 8	10) 37	16) 52
5) 43	11) 39	17) 32
6) 69	12) 27	18) 22

Adding three–digit numbers

1) 290	7) 846	14) 540
2) 657	8) 1,203	15) 809
3) 512	9) 1,045	16) 843
4) 580	10) 347	17) 834
5) 862	11) 273	18) 976
6) 385	12) 468	
	13) 470	

Adding hundreds

1) 200	10) 1,100	19) 400
2) 300	11) 1,000	20) 1,000
3) 400	12) 900	21) 800
4) 500	13) 1,000	22) 700
5) 400	14) 800	23) 1,200
6) 600	15) 600	24) 1,400
7) 500	16) 1,000	25) 1,100
8) 700	17) 900	
9) 800	18) 800	

Adding 4–digit numbers

1)	7,845	7)	4,364	13)	1,300
2)	6,486	8)	8,085	14)	530
3)	8,468	9)	9,457	15)	1,560
4)	5,801	10)	131	16)	$9,355
5)	9,451	11)	1,500		
6)	8,404	12)	1,100		

Subtracting 4–digit numbers

1)	1,118	7)	2,044	13)	1,050
2)	1,135	8)	5,461	14)	1,890
3)	2,298	9)	901	15)	320
4)	3,097	10)	1,103	16)	1,792
5)	1,891	11)	1,329		
6)	4,609	12)	546		

Chapter 3: Multiplication and Division

Topics that you'll practice in this chapter:

✓ Multiplication

✓ Division

✓ Long Division by One Digit

✓ Division with Remainders

Multiplication

✍ *Find the answers.*

1) $\begin{array}{r} 45 \\ \times\ 13 \\ \hline \\ \hline \end{array}$

2) $\begin{array}{r} 32 \\ \times\ 10 \\ \hline \\ \hline \end{array}$

3) $\begin{array}{r} 19 \\ \times\ 12 \\ \hline \\ \hline \end{array}$

4) $\begin{array}{r} 25 \\ \times\ 15 \\ \hline \\ \hline \end{array}$

5) $\begin{array}{r} 38 \\ \times\ 14 \\ \hline \\ \hline \end{array}$

6) $\begin{array}{r} 34 \\ \times\ 24 \\ \hline \\ \hline \end{array}$

7) $\begin{array}{r} 52 \\ \times\ 11 \\ \hline \\ \hline \end{array}$

8) $\begin{array}{r} 47 \\ \times\ 20 \\ \hline \\ \hline \end{array}$

9) $\begin{array}{r} 120 \\ \times\ 9 \\ \hline \\ \hline \end{array}$

10) $\begin{array}{r} 563 \\ \times\ 4 \\ \hline \\ \hline \end{array}$

11) $\begin{array}{r} 365 \\ \times\ 5 \\ \hline \\ \hline \end{array}$

12) $\begin{array}{r} 89 \\ \times\ 25 \\ \hline \\ \hline \end{array}$

13) $\begin{array}{r} 478 \\ \times\ 34 \\ \hline \\ \hline \end{array}$

14) $\begin{array}{r} 956 \\ \times\ 26 \\ \hline \\ \hline \end{array}$

15) $\begin{array}{r} 391 \\ \times\ 78 \\ \hline \\ \hline \end{array}$

16) The Haunted House Ride runs 5 times a day. It has 6 cars, each of which can hold 4 people. How many people can ride the Haunted House Ride in one day?

17) Each train car has 3 rows of seats. There are 4 seats in each row. How many seats are there in 5 train cars?

Division

✎ *Find each missing number.*

1) $8 \div \underline{\quad} = 4$

2) $\underline{\quad} \div 4 = 3$

3) $14 \div \underline{\quad} = 2$

4) $\underline{\quad} \div 5 = 3$

5) $18 \div \underline{\quad} = 2$

6) $\underline{\quad} \div 7 = 3$

7) $10 \div \underline{\quad} = 1$

8) $48 \div 12 = \underline{\quad}$

9) $99 \div \underline{\quad} = 9$

10) $70 \div 10 = \underline{\quad}$

11) $44 \div \underline{\quad} = 4$

12) $24 \div \underline{\quad} = 2$

13) $\underline{\quad} \div 10 = 4$

14) $110 \div 11 = \underline{\quad}$

15) $12 \div \underline{\quad} = 1$

16) $90 \div \underline{\quad} = 9$

17) $\underline{\quad} \div 11 = 8$

18) $\underline{\quad} \div 12 = 11$

19) $60 \div \underline{\quad} = 6$

20) $\underline{\quad} \div 11 = 12$

21) $84 \div 12 = \underline{\quad}$

22) $80 \div 10 = \underline{\quad}$

23) $11 \div 11 = \underline{\quad}$

24) $144 \div \underline{\quad} = 12$

25) Anna has 120 books. She wants to put them in equal numbers on 12 bookshelves. How many books can she put on a bookshelf?   _____ books

26) If dividend is 99 and the quotient is 11, then what is the divisor? _____

Long Division by One Digit

✎ *Find the quotient.*

1) $8\overline{)40}$ =

2) $5\overline{)30}$ =

3) $6\overline{)36}$ =

4) $4\overline{)40}$ =

5) $6\overline{)42}$ =

6) $8\overline{)64}$ =

7) $7\overline{)35}$ =

8) $7\overline{)49}$ =

9) $8\overline{)56}$ =

10) $9\overline{)36}$ =

11) $7\overline{)28}$ =

12) $8\overline{)32}$ =

13) $8\overline{)72}$ =

14) $7\overline{)70}$ =

15) $6\overline{)54}$ =

16) $11\overline{)99}$ =

17) $12\overline{)144}$ =

18) $5\overline{)60}$ =

19) $6\overline{)84}$ =

20) $7\overline{)112}$ =

21) $8\overline{)152}$ =

22) $8\overline{)168}$ =

23) $7\overline{)819}$ =

24) $5\overline{)225}$ =

25) $6\overline{)792}$ =

26) $5\overline{)350}$ =

27) $6\overline{)174}$ =

28) $8\overline{)104}$ =

29) $3\overline{)102}$ =

30) $9\overline{)189}$ =

31) $5\overline{)115}$ =

32) $2\overline{)120}$ =

33) $7\overline{)112}$ =

34) $4\overline{)148}$ =

35) $9\overline{)126}$ =

36) $6\overline{)240}$ =

37) $4\overline{)576}$ =

38) $4\overline{)512}$ =

39) $9\overline{)1278}$ =

40) $8\overline{)2768}$ =

41) $6\overline{)1224}$ =

42) $4\overline{)3412}$ =

Division with Remainders

✐ *Find the quotient with remainder.*

1) $5\overline{)27}$

2) $2\overline{)19}$

3) $4\overline{)17}$

4) $7\overline{)23}$

5) $6\overline{)34}$

6) $5\overline{)41}$

7) $5\overline{)26}$

8) $7\overline{)29}$

9) $4\overline{)33}$

10) $7\overline{)46}$

11) $8\overline{)59}$

12) $8\overline{)67}$

13) $9\overline{)65}$

14) $7\overline{)50}$

15) $9\overline{)84}$

16) $9\overline{)95}$

17) $4\overline{)85}$

18) $7\overline{)93}$

19) $8\overline{)117}$

20) $5\overline{)124}$

21) $8\overline{)189}$

22) $7\overline{)256}$

23) $4\overline{)265}$

24) $6\overline{)232}$

25) $5\overline{)592}$

26) $3\overline{)295}$

27) $6\overline{)553}$

28) $5\overline{)214}$

29) $3\overline{)440}$

30) $7\overline{)673}$

31) $4\overline{)213}$

32) $2\overline{)820}$

33) $5\overline{)496}$

34) $6\overline{)791}$

35) $4\overline{)647}$

36) $7\overline{)780}$

37) $4\overline{)5910}$

38) $8\overline{)3515}$

39) $7\overline{)2355}$

40) $9\overline{)1232}$

41) $8\overline{)6029}$

42) $4\overline{)6743}$

Answers of Worksheets – Chapter 3

Multiplication

1)	585	7)	572	13)	16,252
2)	320	8)	940	14)	24,856
3)	228	9)	1,080	15)	30,498
4)	375	10)	2,252	16)	120
5)	532	11)	1,825	17)	60
6)	816	12)	2,225		

Division

1)	2	10)	7	19)	10
2)	12	11)	11	20)	132
3)	7	12)	12	21)	7
4)	15	13)	40	22)	8
5)	9	14)	10	23)	1
6)	21	15)	12	24)	12
7)	10	16)	10	25)	10
8)	4	17)	88	26)	9
9)	11	18)	132		

Long Division by One Digit

1)	5	15)	9	29)	34
2)	6	16)	9	30)	21
3)	6	17)	12	31)	23
4)	10	18)	12	32)	60
5)	7	19)	14	33)	16
6)	8	20)	16	34)	37
7)	5	21)	19	35)	14
8)	7	22)	21	36)	40
9)	7	23)	117	37)	144
10)	4	24)	45	38)	128
11)	9	25)	132	39)	142
12)	4	26)	70	40)	346
13)	9	27)	29	41)	204
14)	10	28)	13	42)	853

Division with Remainders

1) 5 *R*2
2) 9 *R*1
3) 4 *R*1
4) 3 *R*2
5) 5 *R*4
6) 8 *R*1
7) 5 *R*1
8) 4 *R*1
9) 8 *R*1
10) 6 *R*4
11) 7 *R*3
12) 8 *R*3
13) 7 *R*2
14) 7 *R*1

15) 9 *R*3
16) 9 *R*5
17) 21 *R*1
18) 13 *R*2
19) 14 *R*5
20) 24 *R*4
21) 23 *R*5
22) 36 *R*4
23) 66 *R*1
24) 38 *R*4
25) 118 *R*4
26) 98 *R*1
27) 92 *R*1
28) 42 *R*4

29) 146 *R*2
30) 96 *R*1
31) 53 *R*1
32) 410 *R*0
33) 99 *R*1
34) 131 *R*5
35) 161 *R*3
36) 111 *R*3
37) 1477 *R*2
38) 439 *R*3
39) 336 *R*3
40) 135 *R*8
41) 753 *R*5
42) 1,685 *R*3

Chapter 4:
Mixed operations

Topics that you'll practice in this chapter:

- ✓ Rounding and Estimating
- ✓ Estimate Sums
- ✓ Estimate Differences
- ✓ Estimate Products
- ✓ Missing Numbers

Rounding and Estimating

 Estimate the sum by rounding each number to the nearest ten.

1) $14 + 68 =$

2) $82 + 12 =$

3) $43 + 66 =$

4) $47 + 65 =$

5) $553 + 232 =$

6) $418 + 846 =$

7) $582 + 277 =$

8) $2771 + 1651 =$

 Estimate the product by rounding each number to the nearest ten.

9) $55 \times 62 =$

10) $14 \times 27 =$

11) $34 \times 66 =$

12) $18 \times 12 =$

13) $62 \times 53 =$

14) $41 \times 26 =$

15) $19 \times 33 =$

16) $76 \times 45 =$

Estimate the sum or product by rounding each number to the nearest ten.

17)
$$\begin{array}{r} 34 \\ \times\ 26 \\ \hline \end{array}$$

19)
$$\begin{array}{r} 78 \\ +\ 92 \\ \hline \end{array}$$

21)
$$\begin{array}{r} 73 \\ \times\ 12 \\ \hline \end{array}$$

18)
$$\begin{array}{r} 53 \\ \times\ 18 \\ \hline \end{array}$$

20)
$$\begin{array}{r} 55 \\ +94 \\ \hline \end{array}$$

22)
$$\begin{array}{r} 81 \\ +\ 53 \\ \hline \end{array}$$

Estimate Sums

✍ *Estimate the sum by rounding each added to the nearest ten.*

1) $55 + 9 =$

2) $13 + 74 =$

3) $83 + 7 =$

4) $32 + 37 =$

5) $13 + 74 =$

6) $34 + 11 =$

7) $39 + 77 =$

8) $25 + 4 =$

9) $61 + 73 =$

10) $64 + 59 =$

11) $14 + 68 =$

12) $82 + 12 =$

13) $43 + 66 =$

14) $45 + 65 =$

15) $96 + 94 =$

16) $29 + 89 =$

17) $78 + 74 =$

18) $39 + 27 =$

19) $91 + 68 =$

20) $48 + 81 =$

21) $14 + 96 =$

22) $52 + 59 =$

23) $553 + 232 =$

24) $52 + 67 =$

Estimate Differences

✎ *Estimate the difference by rounding each number to the nearest ten.*

1) $46 - 11 =$

2) $23 - 14 =$

3) $68 - 36 =$

4) $22 - 13 =$

5) $59 - 36 =$

6) $34 - 11 =$

7) $67 - 37 =$

8) $38 - 19 =$

9) $84 - 38 =$

10) $68 - 48 =$

11) $58 - 16 =$

12) $72 - 27 =$

13) $63 - 33 =$

14) $49 - 32 =$

15) $94 - 63 =$

16) $55 - 32 =$

17) $87 - 74 =$

18) $32 - 11 =$

19) $46 - 39 =$

20) $99 - 36 =$

21) $94 - 78 =$

22) $75 - 23 =$

23) $99 - 19 =$

24) $86 - 43 =$

Estimate Products

✍ *Estimate the products.*

1) $27 \times 18 =$

2) $13 \times 17 =$

3) $43 \times 19 =$

4) $22 \times 25 =$

5) $68 \times 23 =$

6) $36 \times 91 =$

7) $53 \times 92 =$

8) $18 \times 38 =$

9) $21 \times 14 =$

10) $83 \times 42 =$

11) $51 \times 32 =$

12) $68 \times 12 =$

13) $47 \times 23 =$

14) $71 \times 58 =$

15) $54 \times 89 =$

16) $37 \times 72 =$

17) $36 \times 93 =$

18) $32 \times 29 =$

19) $41 \times 37 =$

20) $54 \times 93 =$

21) $89 \times 72 =$

22) $77 \times 22 =$

23) $53 \times 13 =$

24) $98 \times 63 =$

Missing Numbers

✎ *Find the missing numbers.*

1) $20 \times \_\_ = 60$

2) $16 \times \_\_ = 32$

3) $\_\_ \times 14 = 84$

4) $16 \times \_\_ = 80$

5) $\_\_ \times 19 = 38$

6) $17 \times \_\_ = 34$

7) $\_\_ \times 1 = 18$

8) $21 \times \_\_ = 42$

9) $20 \times \_\_ = 80$

10) $15 \times 7 = \_\_$

11) $18 \times 9 = \_\_$

12) $21 \times 4 = \_\_$

13) $23 \times 7 = \_\_$

14) $\_\_ \times 25 = 75$

15) $24 \times \_\_ = 120$

16) $22 \times 4 = \_\_$

17) $20 \times \_\_ = 140$

18) $17 \times \_\_ = 153$

19) $\_\_ \times 15 = 120$

20) $21 \times 6 = \_\_$

21) $\_\_ \times 22 = 154$

22) $19 \times \_\_ = 76$

23) $23 \times 9 = \_\_$

24) $25 \times 6 = \_\_$

25) $\_\_ \times 18 = 36$

26) $24 \times \_\_ = 48$

Answers of Worksheets – Chapter 4

Rounding and Estimating

1) 80	9) 3,600	17) 900
2) 90	10) 300	18) 1,000
3) 110	11) 2,100	19) 170
4) 120	12) 200	20) 150
5) 780	13) 3,000	21) 700
6) 1,270	14) 1,200	22) 130
7) 860	15) 600	
8) 4,420	16) 4,000	

Estimate sums

1) 70	9) 130	17) 150
2) 80	10) 120	18) 70
3) 90	11) 80	19) 160
4) 70	12) 90	20) 130
5) 80	13) 110	21) 110
6) 40	14) 120	22) 110
7) 120	15) 190	23) 780
8) 30	16) 120	24) 120

Estimate differences

1) 40	9) 40	17) 20
2) 10	10) 20	18) 20
3) 30	11) 40	19) 10
4) 10	12) 40	20) 60
5) 20	13) 30	21) 10
6) 20	14) 20	22) 60
7) 30	15) 30	23) 80
8) 20	16) 30	24) 50

Estimate products

1) 600	7) 4500	13) 1000
2) 200	8) 800	14) 4200
3) 600	9) 200	15) 4500
4) 800	10) 3200	16) 2800
5) 1400	11) 1500	17) 3600
6) 3600	12) 700	18) 900

19) 1600   21) 6300   23) 500
20) 4500   22) 1600   24) 6000

Missing Numbers

1) 3   10) 105   19) 8
2) 2   11) 162   20) 126
3) 6   12) 84   21) 7
4) 5   13) 161   22) 4
5) 2   14) 3   23) 207
6) 2   15) 5   24) 150
7) 18   16) 88   25) 2
8) 2   17) 7   26) 2
9) 4   18) 9

Chapter 5:
Number Theory

Topics that you'll practice in this chapter:

- ✓ Factoring Numbers
- ✓ Prime Factorization
- ✓ Greatest Common Factor
- ✓ Least Common Multiple
- ✓ Divisibility Rules

Factoring Numbers

✍ *List all positive factors of each number.*

1) 8

2) 9

3) 15

4) 16

5) 25

6) 28

7) 26

8) 35

9) 42

10) 48

11) 50

12) 36

13) 55

14) 40

15) 62

16) 84

17) 75

18) 68

19) 96

20) 78

21) 94

22) 82

23) 81

24) 72

Prime Factorization

✎ *Factor the following numbers to their prime factors.*

1) 20	9) 62	17) 18
2) 12	10) 49	18) 32
3) 16	11) 51	19) 15
4) 27	12) 78	20) 33
5) 36	13) 63	21) 40
6) 42	14) 77	22) 50
7) 58	15) 46	23) 45
8) 35	16) 69	24) 70

Greatest Common Factor

✎ *Find the GCF for each number pair.*

1) 4, 2

2) 3, 5

3) 2, 6

4) 4, 7

5) 5, 10

6) 6, 12

7) 7, 14

8) 6, 14

9) 5, 12

10) 4, 14

11) 15, 18

12) 12, 20

13) 12, 16

14) 15, 27

15) 8, 24

16) 28, 16

17) 32, 24

18) 18, 36

19) 26, 20

20) 30, 14

21) 24, 20

22) 14, 22

23) 25, 15

24) 28, 32

Least Common Multiple

✍ *Find the LCM for each number pair.*

1) 3, 6

2) 5, 10

3) 6, 14

4) 8, 9

5) 6, 18

6) 10, 12

7) 4, 12

8) 5, 15

9) 4, 18

10) 9, 12

11) 12, 16

12) 15, 18

13) 8, 24

14) 9, 28

15) 12, 24

16) 15, 20

17) 25, 18

18) 27, 24

19) 28, 18

20) 16, 30

21) 14, 28

22) 20, 35

23) 25, 30

24) 32, 27

Divisibility Rules

✎ *Use the divisibility rules to underline the factors of the number.*

8	<u>2</u> 3 <u>4</u> 5 6 7 <u>8</u> 9 10
1) 16	2 3 4 5 6 7 8 9 10
2) 10	2 3 4 5 6 7 8 9 10
3) 15	2 3 4 5 6 7 8 9 10
4) 28	2 3 4 5 6 7 8 9 10
5) 36	2 3 4 5 6 7 8 9 10
6) 15	2 3 4 5 6 7 8 9 10
7) 27	2 3 4 5 6 7 8 9 10
8) 70	2 3 4 5 6 7 8 9 10
9) 57	2 3 4 5 6 7 8 9 10
10) 102	2 3 4 5 6 7 8 9 10
11) 144	2 3 4 5 6 7 8 9 10
12) 75	2 3 4 5 6 7 8 9 10

Answers of Worksheets – Chapter 5

Factoring Numbers

1) 1, 2, 4, 8
2) 1, 3, 9
3) 1, 3, 5, 15
4) 1, 2, 4, 8, 16
5) 1, 5, 25
6) 1, 2, 4, 7, 14, 28
7) 1, 2, 13, 26
8) 1, 5, 7, 35
9) 1, 2, 3, 6, 7, 14, 21, 42
10) 1, 2, 3, 4, 6, 8, 12, 16, 24, 48
11) 1, 2, 5, 10, 25, 50
12) 1, 2, 3, 4, 6, 9, 12, 18, 36

13) 1, 5, 11, 55
14) 1, 2, 4, 5, 8, 10, 20, 40
15) 1, 2, 31, 62
16) 1, 2, 3, 4, 6, 7, 12, 14, 21, 28, 42, 84
17) 1, 3, 5, 15, 25, 75
18) 1, 2, 4, 17, 34, 68
19) 1, 2, 3, 4, 6, 8, 12, 16, 24, 32, 48, 96
20) 1, 2, 3, 6, 13, 26, 39, 78
21) 1, 2, 47, 94
22) 1, 2, 41, 82
23) 1, 3, 9, 27, 81
24) 1, 2, 3, 4, 6, 8, 9, 12, 18, 24, 36, 72

Prime Factorization

1) 2 . 2 . 5
2) 2 . 2 . 3
3) 2 . 2 . 2 . 2
4) 3 . 3 . 3
5) 2 . 2 . 3 . 3
6) 2 . 3 . 7
7) 2 . 29
8) 5 . 7

9) 2 . 31
10) 7 . 7
11) 3 . 17
12) 2 . 3 . 13
13) 3 . 3 . 7
14) 7 . 11
15) 2 . 23
16) 3 . 23

17) 2 . 3 . 3
18) 2 . 2 . 2 . 2 . 2
19) 3 . 5
20) 3 . 11
21) 2 . 2 . 2 . 5
22) 2 . 5 . 5
23) 3 . 3 . 5
24) 2 . 5 . 7

Greatest Common Factor

1) 2
2) 1
3) 2
4) 1
5) 5
6) 6
7) 7
8) 2

9) 1
10) 2
11) 3
12) 4
13) 4
14) 3
15) 8
16) 4

17) 8
18) 18
19) 2
20) 2
21) 4
22) 2
23) 5
24) 4

Least Common Multiple

1) 6
2) 10

3) 42
4) 72

5) 18
6) 60

7) 12	13) 24	19) 252
8) 15	14) 252	20) 240
9) 36	15) 24	21) 28
10) 36	16) 60	22) 140
11) 48	17) 450	23) 150
12) 90	18) 216	24) 864

Divisibility Rules

8

1) 16	2 3 **4** 5 6 7 **8** 9 10
2) 10	**2** 3 **4** 5 6 7 **8** 9 10
3) 15	**2** 3 4 **5** 6 7 8 9 **10**
4) 28	2 **3** 4 **5** 6 7 8 9 10
5) 36	**2** 3 **4** 5 6 **7** 8 9 10
6) 18	**2** 3 **4** 5 **6** 7 8 **9** 10
7) 27	**2** **3** 4 5 **6** 7 8 **9** 10
8) 70	**2** **3** 4 5 6 7 8 **9** 10
9) 57	**2** 3 4 **5** 6 **7** 8 9 **10**
10) 102	**2** **3** 4 5 6 7 8 9 10
11) 144	**2** **3** 4 5 **6** 7 8 9 10
12) 75	**2** **3** **4** 5 **6** 7 **8** **9** 10
	2 **3** 4 **5** 6 7 8 9 10

Chapter 6:
Data and Graphs

Topics that you'll practice in this chapter:

- ✓ Graph Points on a Coordinate Plane
- ✓ Bar Graph
- ✓ Tally and Pictographs
- ✓ Line Graphs
- ✓ Stem–And–Leaf Plot
- ✓ Scatter Plots

Graph Points on a Coordinate Plane

✍ *Plot each point on the coordinate grid.*

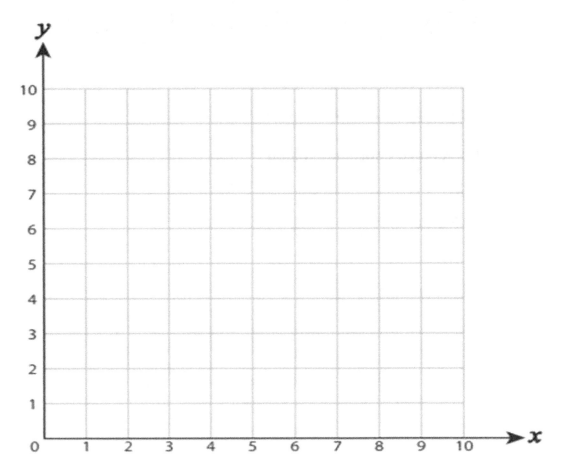

1) *A* (3, 6)

2) *B* (1, 3)

3) *C* (3, 7)

4) *D* (8, 6)

5) *E* (5, 2)

6) *F* (9, 3)

7) *G* (2, 1)

8) *H* (4, 2)

9) *I* (6, 6)

10) *J* (7, 2)

11) *K* (8, 3)

12) *L* (2, 9)

Bar Graph

✎ *Graph the given information as a bar graph.*

Day	Hot dogs sold
Monday	90
Tuesday	70
Wednesday	30
Thursday	20
Friday	60

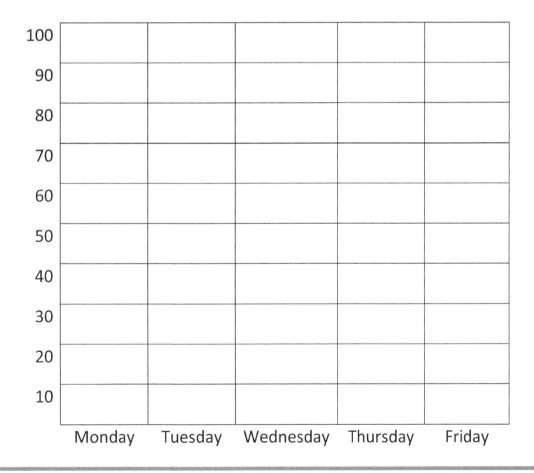

Tally and Pictographs

✎ *Using the key, draw the pictograph to show the information.*

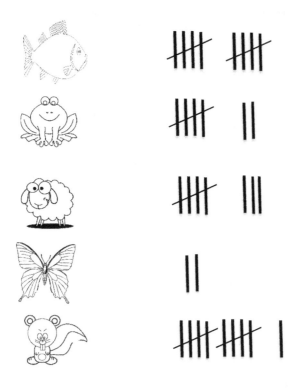

Key: ☺ = 2 animals

Line Graphs

✎*David work as a salesman in a store. He records the number of shoes sold in five days on a line graph. Use the graph to answer the questions.*

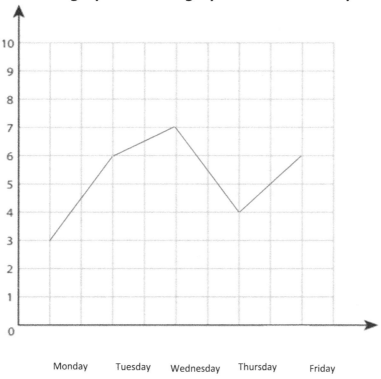

1) How many cars were sold on Monday?

2) Which day had the minimum sales of shoes?

3) Which day had the maximum number of shoes sold?

4) How many shoes were sold in 5 days?

Histograms

✎ *Use the following Graph to complete the table.*

Day	Distance (km)
1	
2	

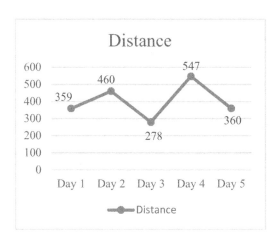

The following table shows the number of births in the US from 2007 to 2012 (in millions).

Year	Number of births (in millions)
2007	4.32
2008	4.25
2009	4.13
2010	4
2011	3.95
2012	3.95

Draw a histogram for the table.

Answers of Worksheets – Chapter 6

Graph Points on a Coordinate Plane

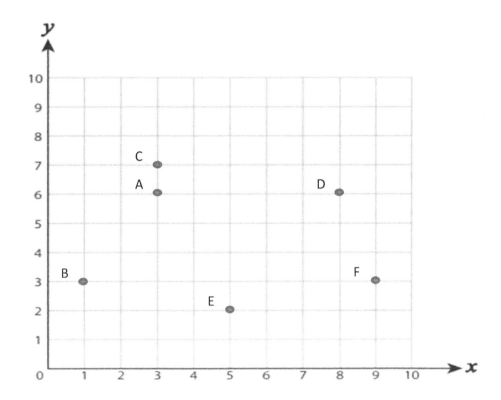

Bar Graph

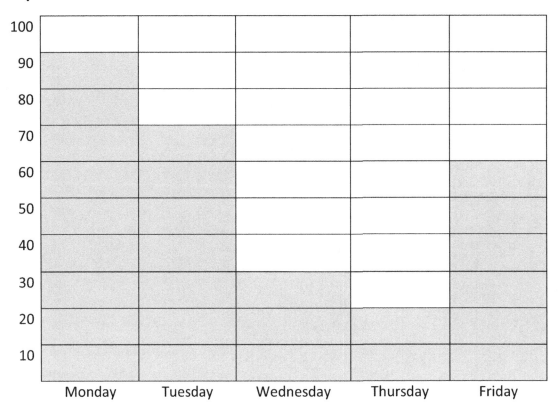

Tally and Pictographs

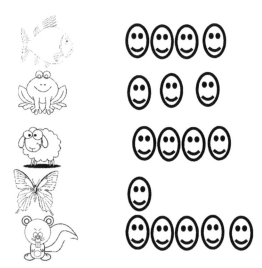

Line Graphs

1) 3
2) Thursday

3) Wednesday
4) 26

Histograms

Day	Distance (km)
1	359
2	460
3	278
4	547
5	360

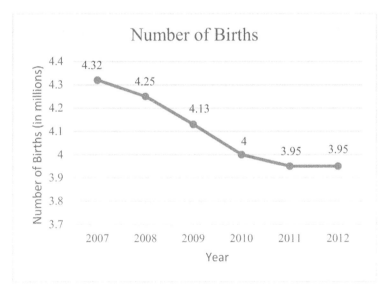

Chapter 7:
Patterns and Sequences

Topics that you'll practice in this chapter:

- ✓ Repeating pattern
- ✓ Growing Patterns
- ✓ Patterns: Numbers

Repeating Pattern

✍ *Circle the picture that comes next in each picture pattern.*

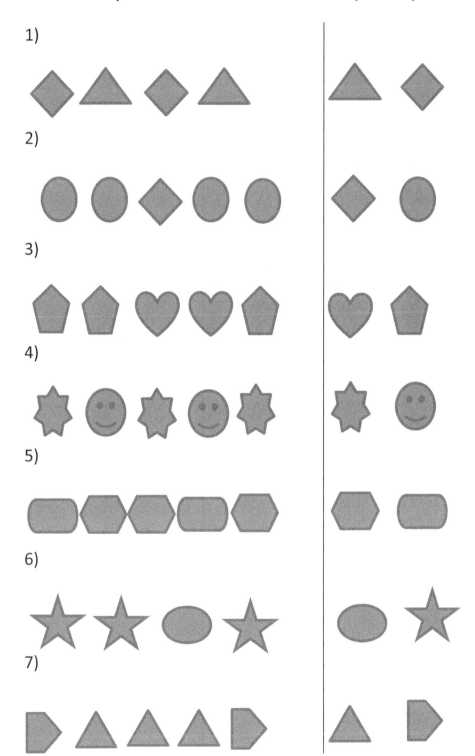

1)

2)

3)

4)

5)

6)

7)

Growing Patterns

✎ *Draw the picture that comes next in each growing pattern.*

1)

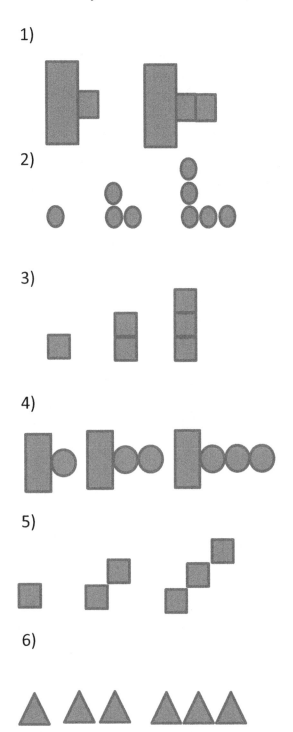

2)

3)

4)

5)

6)

Patterns: Numbers

✎ *Write the numbers that come next.*

1) 12, 14, 16, 18,   ____, ____, ____, ____

2) 7, 14, 21, 28,   ____, ____, ____, ____

3) 15, 25, 35, 45,   ____, ____, ____, ____

4) 11, 22, 33, 44,   ____, ____, ____, ____

5) 10, 18, 26, 34, 42,   ____, ____, ____, ____

6) 61, 55, 49, 43, 37,   ____, ____, ____, ____

7) 45, 56, 67, 78,   ____, ____, ____, ____

✎ *Write the next three numbers in each counting sequence.*

8) −32, −23, −14, ____, ____, ____, ____

9) 543, 528, 513, ____, ____, ____, ____

10) ____, ____, 56, 64, ____, 80

11) 23, 34, ____, ____, 67, ____

12) 24, 31, ____, ____, ____

13) 52, 45, ____, ____, ____

14) 51, 44, 37, ____, ____, ____

15) 64, 51, 38, ____, ____, ____

Answers of Worksheets – Chapter 7

Repeating pattern

1)

2)

3)

4)

5)

6)

7)

Growing patterns

1)

2)

3)

4)

5)

6)

Patterns: Numbers

1) 12, 14, 16, 18, 20, 22, 24, 26
2) 7, 14, 21, 28, 35, 42, 49, 56
3) 15, 25, 35, 45, 55, 65, 75, 85
4) 11, 22, 33, 44, 55, 66, 77, 88
5) 10, 18, 26, 34, 42, 50, 58, 66
6) 61, 55, 49, 43, 37, 31, 25, 19
7) 45, 56, 67, 78, 89, 100, 111, 122
8) − 5, 4, 13, 22
9) 498, 483, 468
10) 40 − 48 − 56 − 64 − 72 − 80
11) 23 − 34 − 45 − 56 − 67 − 78
12) 38 − 45 − 52
13) 38 − 31 − 24
14) 30, 23, 16
15) 25, 12, − 1

Chapter 8: Money

Topics that you'll practice in this chapter:

- ✓ Add Money Amounts
- ✓ Subtract Money Amounts
- ✓ Money: Word Problems

Add Money Amounts

✎ *Add.*

1)
$$
\begin{array}{r}
\$314 \\
+\$152 \\
\hline
\end{array}
\qquad
\begin{array}{r}
\$624 \\
+\$410 \\
\hline
\end{array}
\qquad
\begin{array}{r}
\$390 \\
+\$215 \\
\hline
\end{array}
$$

2)
$$
\begin{array}{r}
\$321 \\
+\$430 \\
\hline
\end{array}
\qquad
\begin{array}{r}
\$530 \\
+\$321 \\
\hline
\end{array}
\qquad
\begin{array}{r}
\$712 \\
+\$145 \\
\hline
\end{array}
$$

3)
$$
\begin{array}{r}
\$411 \\
+\$316 \\
\hline
\end{array}
\qquad
\begin{array}{r}
\$559 \\
+\$228 \\
\hline
\end{array}
\qquad
\begin{array}{r}
\$731 \\
+\$213 \\
\hline
\end{array}
$$

4)
$$
\begin{array}{r}
\$621 \\
+\$168 \\
\hline
\end{array}
\qquad
\begin{array}{r}
\$321 \\
+\$129 \\
\hline
\end{array}
\qquad
\begin{array}{r}
\$615 \\
+\$371 \\
\hline
\end{array}
$$

5)
$$
\begin{array}{r}
\$526 \\
+\$228 \\
\hline
\end{array}
\qquad
\begin{array}{r}
\$287 \\
+\$129 \\
\hline
\end{array}
\qquad
\begin{array}{r}
\$493 \\
+\$274 \\
\hline
\end{array}
$$

6)
$$
\begin{array}{r}
\$386 \\
+\$464 \\
\hline
\end{array}
\qquad
\begin{array}{r}
\$275 \\
+\$175 \\
\hline
\end{array}
\qquad
\begin{array}{r}
\$636 \\
+\$295 \\
\hline
\end{array}
$$

7)
$$
\begin{array}{r}
\$489 \\
+\ \$378 \\
\hline
\end{array}
\qquad
\begin{array}{r}
\$579 \\
+\$459 \\
\hline
\end{array}
\qquad
\begin{array}{r}
\$737 \\
+\$462 \\
\hline
\end{array}
$$

Subtract Money Amounts

✍ *Subtract.*

1)
$825
−$166
_____

$651
−$110
_____

$754
−$565
_____

2)
$539
−$137
_____

$498
−$359
_____

$992
−$549
_____

3)
$436
−$219
_____

$512
−$128
_____

$632
−$444
_____

4)
$345
−$127
_____

$419
−$361
_____

$397
−$231
_____

5)
$452
−$298
_____

$583
−$362
_____

$684
−$495
_____

6)
$735
−$599
_____

$829
−$714
_____

$984
−$582
_____

7) Linda had $120. She bought some game tickets for $70. How much did she have left?

Money: Word Problems

✍ *Solve.*

1) How many boxes of envelopes can you buy with $18 if one box costs $3?

2) After paying $6.25 for a salad, Ella has $45.56. How much money did she have before buying the salad?

3) How many packages of diapers can you buy with $50 if one package costs $5?

4) Last week James ran 20 miles more than Michael. James ran 56 miles. How many miles did Michael run?

5) Last Friday Jacob had $32.52. Over the weekend he received some money for cleaning the attic. He now has $44. How much money did he receive?

6) After paying $10.12 for a sandwich, Amelia has $35.50. How much money did she have before buying the sandwich?

Answers of Worksheets – Chapter 8

Add Money Amounts

1) $466, 1,034, 605$
2) $751, 851, 857$
3) $727, 787, 944$
4) $789, 450, 986$
5) $754, 416, 767$
6) $850, 450, 931$
7) $867, 1,038, 1,199$

Subtract Money Amounts

1) $659 - 541 - 189$
2) $402 - 139 - 443$
3) $217 - 384 - 188$
4) $218, 58, 166$
5) $154, 221, 189$
6) $136, 115, 402$
7) 50

Money: word problem

1) 6
2) 51.81
3) 10
4) 36
5) 11.48
6) 45.62

Chapter 9: Measurement

Topics that you'll practice in this chapter:

- ✓ Convert Measurement Units
- ✓ Metric units
- ✓ Distance Measurement
- ✓ Weight Measurement

Convert Measurement Units

✎ *Convert to an appropriate measurement unit. (Round to the nearest Hundredths)*

1) $4\,m = \underline{\hspace{1cm}} cm$

2) $50\,cm = \underline{\hspace{1cm}} m$

3) $5\,m = \underline{\hspace{1cm}} cm$

4) $3\,feet = \underline{\hspace{1cm}} inches$

5) $5\,feet = \underline{\hspace{1cm}} cm$

6) $2\,feet = \underline{\hspace{1cm}} inches$

7) $1\,inch = \underline{\hspace{1cm}} cm$

8) $4\,feet = \underline{\hspace{1cm}} inches$

9) $8\,inches = \underline{\hspace{1cm}} foot$

10) $10\,feet = \underline{\hspace{1cm}} m$

11) $15\,cm = \underline{\hspace{1cm}} m$

12) $5\,inches = \underline{\hspace{1cm}} cm$

13) $10\,inches = \underline{\hspace{1cm}} m$

14) $15\,inches = \underline{\hspace{1cm}} cm$

15) $12\,inches = \underline{\hspace{1cm}} m$

16) $8\,feet = \underline{\hspace{1cm}} inches$

17) $25\,cm = \underline{\hspace{1cm}} inches$

18) $11\,inches = \underline{\hspace{1cm}} cm$

19) $1\,m = \underline{\hspace{1cm}} inches$

20) $80\,inches = \underline{\hspace{1cm}} m$

21) $200\,cm = \underline{\hspace{1cm}} m$

22) $5\,m = \underline{\hspace{1cm}} cm$

23) $12\,feet = \underline{\hspace{1cm}} inches$

24) $10\,yards = \underline{\hspace{1cm}} inches$

25) $16\,feet = \underline{\hspace{1cm}} inches$

26) $48\,inches = \underline{\hspace{1cm}} Feet$

27) $4\,inches = \underline{\hspace{1cm}} cm$

28) $12.5\,cm = \underline{\hspace{1cm}} inches$

29) $6\,feet = \underline{\hspace{1cm}} inches$

30) $10\,feet = \underline{\hspace{1cm}} inches$

31) $12\,yards = \underline{\hspace{1cm}} feet$

32) $7\,yards = \underline{\hspace{1cm}} feet$

Metric Units

✎ *Convert to an appropriate Metric unit.*

1) $1 \, cm = \underline{\hspace{1cm}} mm$

2) $1 \, m = \underline{\hspace{1cm}} mm$

3) $5 \, cm = \underline{\hspace{1cm}} mm$

4) $0.1 \, cm = \underline{\hspace{1cm}} mm$

5) $0.2 \, m = \underline{\hspace{1cm}} cm$

6) $10 \, mm = \underline{\hspace{1cm}} cm$

7) $50 \, mm = \underline{\hspace{1cm}} m$

8) $10 \, cm = \underline{\hspace{1cm}} m$

9) $100 \, mm = \underline{\hspace{1cm}} cm$

10) $0.05 \, m = \underline{\hspace{1cm}} mm$

11) $1 \, km = \underline{\hspace{1cm}} m$

12) $0.01 \, km = \underline{\hspace{1cm}} m$

13) $500 \, cm = \underline{\hspace{1cm}} m$

14) $0.50 \, km \underline{\hspace{1cm}} m$

15) $100 \, cm = \underline{\hspace{1cm}} m$

16) $80 \, cm = \underline{\hspace{1cm}} mm$

17) $4 \, mm = \underline{\hspace{1cm}} cm$

18) $0.6 \, m = \underline{\hspace{1cm}} mm$

19) $2 \, m = \underline{\hspace{1cm}} cm$

20) $0.03 \, km = \underline{\hspace{1cm}} m$

21) $3000 \, mm = \underline{\hspace{1cm}} km$

22) $5 \, cm = \underline{\hspace{1cm}} m$

23) $0.03 \, m = \underline{\hspace{1cm}} cm$

24) $1000 \, mm = \underline{\hspace{1cm}} km$

25) $600 \, mm = \underline{\hspace{1cm}} m$

26) $0.77 \, km = \underline{\hspace{1cm}} mm$

27) $0.08 \, km = \underline{\hspace{1cm}} m$

28) $0.30 \, m = \underline{\hspace{1cm}} cm$

29) $400 \, m = \underline{\hspace{1cm}} km$

30) $5000 \, cm = \underline{\hspace{1cm}} km$

31) $40 \, mm = \underline{\hspace{1cm}} cm$

32) $800 \, m = \underline{\hspace{1cm}} km$

Distance Measurement

✎ **Convert to the new units. (Round to the nearest Hundredths)**

1) $1 \, mi = \underline{\hspace{1cm}} ft$

2) $1 \, mi = \underline{\hspace{1cm}} yd$

3) $1 \, yd = \underline{\hspace{1cm}} m$

4) $2 \, yd = \underline{\hspace{1cm}} ft$

5) $2 \, mi = \underline{\hspace{1cm}} yd$

6) $3 \, mi = \underline{\hspace{1cm}} m$

7) $5 \, mi = \underline{\hspace{1cm}} ft$

8) $6 \, m = \underline{\hspace{1cm}} ft$

9) $4 \, mi = \underline{\hspace{1cm}} m$

10) $10 \, mi = \underline{\hspace{1cm}} yd$

11) $9 \, mi = \underline{\hspace{1cm}} yd$

12) $12 \, mi = \underline{\hspace{1cm}} yd$

13) $10 \, mi = \underline{\hspace{1cm}} ft$

14) $15 \, mi = \underline{\hspace{1cm}} ft$

15) $20 \, mi = \underline{\hspace{1cm}} yd$

16) $16 \, mi = \underline{\hspace{1cm}} yd$

17) $2 \, mi = \underline{\hspace{1cm}} ft$

18) $21 \, mi = \underline{\hspace{1cm}} ft$

19) $6 \, mi = \underline{\hspace{1cm}} ft$

20) $3 \, mi = \underline{\hspace{1cm}} yd$

21) $72 \, mi = \underline{\hspace{1cm}} ft$

22) $41 \, mi = \underline{\hspace{1cm}} yd$

23) $62 \, mi = \underline{\hspace{1cm}} yd$

24) $39 \, mi = \underline{\hspace{1cm}} yd$

25) $7 \, mi = \underline{\hspace{1cm}} yd$

26) $94 \, mi = \underline{\hspace{1cm}} yd$

27) $87 \, mi = \underline{\hspace{1cm}} yd$

28) $23 \, mi = \underline{\hspace{1cm}} yd$

29) $2 \, mi = \underline{\hspace{1cm}} m$

30) $5 \, mi = \underline{\hspace{1cm}} m$

31) $6 \, mi = \underline{\hspace{1cm}} m$

32) $3 \, mi = \underline{\hspace{1cm}} m$

Weight Measurement

✎ *Convert to grams.*

1) $1 \, kg =$ _____ g

2) $3 \, kg =$ _____ g

3) $5 \, kg =$ _____ g

4) $4 \, kg =$ _____ g

5) $0.01 \, kg =$ _____ g

6) $0.2 \, kg =$ _____ g

7) $0.04 \, kg =$ _____ g

8) $0.05 \, kg =$ _____ g

9) $0.5 \, kg =$ _____ g

10) $3.2 \, kg =$ _____ g

11) $8.2 \, kg =$ _____ g

12) $9.2 \, kg =$ _____ g

13) $35 \, kg =$ _____ g

14) $87 \, kg =$ _____ g

15) $45 \, kg =$ _____ g

16) $15 \, kg =$ _____ g

17) $0.32 \, kg =$ _____ g

18) $81 \, kg =$ _____ g

✎ *Convert to kilograms.*

19) $10,000 \, g =$ _____ kg

20) $20,000 \, g =$ _____ kg

21) $3,000 \, g =$ _____ kg

22) $100,000 \, g =$ _____ kg

23) $150,000 \, g =$ _____ kg

24) $120,000 \, g =$ _____ kg

25) $200,000 \, g =$ _____ kg

26) $30,000 \, g =$ _____ kg

27) $800,000 \, g =$ _____ kg

28) $20,000 \, g =$ _____ kg

29) $40,000 \, g =$ _____ kg

30) $500,000 \, g =$ _____ kg

Answers of Worksheets – Chapter 9

Inches & Centimeters

1) $4\ m\ =\ 400\ cm$
2) $50\ cm\ =\ 0.5\ m$
3) $5\ m\ =\ 500\ cm$
4) $3\ feet\ =\ 36\ inches$
5) $5\ feet\ =\ 152.4\ cm$
6) $2\ feet\ =\ 24\ inches$
7) $1\ inch\ =\ 2.54\ cm$
8) $4\ feet\ =\ 48\ inches$
9) $8\ inches\ =\ 0.67\ foot$
10) $10\ feet\ =\ 3.05\ m$
11) $15\ cm\ =\ 0.15\ m$
12) $5\ inches\ =\ 12.7\ cm$
13) $10\ inches\ =\ 0.25\ m$
14) $15\ inches\ =\ 38.1\ cm$
15) $12\ inches\ =\ 0.3\ m$
16) $8\ feet\ =\ 96\ inches$
17) $25\ cm\ =\ 9.84\ inches$
18) $11\ inch\ =\ 27.94\ cm$
19) $1\ m\ =\ 39.37\ inches$
20) $80\ inch\ =\ 2.03\ m$
21) $200\ cm\ =\ 2\ m$
22) $5\ m\ =\ 500\ cm$
23) $12\ feet\ =\ 144\ inches$
24) $10\ yards\ =\ 360\ inches$
25) $16\ feet\ =\ 192\ inches$
26) $48\ inches\ =\ 4\ Feet$
27) $4\ inch\ =\ 10.16\ cm$
28) $12.5\ cm\ =\ 4.92\ inches$
29) $6\ feet\ =\ 72\ inches$
30) $10\ feet\ =\ 120\ inches$
31) $12\ yards\ =\ 36\ feet$
32) $7\ yards\ =\ 21\ feet$

Metric Units

1) $1\ cm\ =\ 10\ mm$
2) $1\ m\ =\ 1000\ mm$
3) $5\ cm\ =\ 50\ mm$
4) $0.1\ cm\ =\ 1\ mm$
5) $0.2\ m\ =\ 20\ cm$
6) $10\ mm\ =\ 1\ cm$
7) $50\ mm\ =\ 0.05\ m$
8) $10\ cm\ =\ 0.10\ m$
9) $100\ mm\ =\ 10\ cm$
10) $0.05\ m\ =\ 50\ mm$
11) $1\ km\ =\ 1,000\ m$
12) $0.01\ km\ =\ 10\ m$
13) $500\ cm\ =\ 5\ m$
14) $0.50\ km\ =\ 500\ m$
15) $100\ cm\ =\ 1\ m$
16) $80\ cm\ =\ 800\ mm$
17) $4\ mm\ =\ 0.4\ cm$
18) $0.6\ m\ =\ 600\ mm$
19) $2\ m\ =\ 200\ cm$
20) $0.03\ km\ =\ 30\ m$
21) $3,000\ mm\ =\ 0.003\ km$
22) $5\ cm\ =\ 0.05\ m$
23) $0.03\ m\ =\ 3\ cm$
24) $1,000\ mm\ =\ 0.001\ km$
25) $600\ mm\ =\ 0.6\ m$
26) $0.77\ km\ =\ 770,000\ mm$
27) $0.08\ km\ =\ 80\ m$
28) $0.30\ m\ =\ 30\ cm$
29) $400\ m\ =\ 0.4\ km$
30) $5,000\ cm\ =\ 0.05\ km$
31) $40\ mm\ =\ 4\ cm$
32) $800\ m\ =\ 0.8\ km$

Distance Measurement

1) $1\ mi = 5{,}280\ ft$
2) $1\ mi = 1{,}760\ yd$
3) $1\ yd = 0.91\ m$
4) $2\ yd = 6\ ft$
5) $2\ mi = 3{,}520\ yd$
6) $3\ mi = 4{,}828\ m$
7) $5\ mi = 26{,}400\ ft$
8) $6\ m = 20\ ft$
9) $4\ mi = 6{,}437\ m$
10) $10\ mi = 17{,}600\ yd$
11) $9\ mi = 15{,}840\ yd$
12) $12\ mi = 21{,}120\ yd$
13) $10\ mi = 52{,}800\ ft$
14) $15\ mi = 79{,}200\ ft$
15) $20\ mi = 35{,}200\ yd$
16) $16\ mi = 28{,}160\ yd$

17) $21\ mi = 110{,}880\ ft$
18) $6\ mi = 31{,}680\ ft$
19) $3\ mi = 5{,}280\ yd$
20) $72\ mi = 380{,}160\ ft$
21) $41\ mi = 72{,}160\ yd$
22) $62\ mi = 109{,}120\ yd$
23) $39\ mi = 68{,}640\ yd$
24) $7\ mi = 12{,}320\ yd$
25) $94\ mi = 165{,}440\ yd$
26) $87\ mi = 153{,}120\ yd$
27) $23\ mi = 40{,}480\ yd$
28) $2\ mi = 3{,}219\ m$
29) $5\ mi = 8{,}047\ m$
30) $6\ mi = 9{,}656\ m$
31) $3\ mi = 4828\ m$

Weight Measurement

1) $1\ kg = 1{,}000\ g$
2) $3\ kg = 3{,}000\ g$
3) $5\ kg = 5{,}000\ g$
4) $4\ kg = 4{,}000\ g$
5) $0.01\ kg = 10\ g$
6) $0.2\ kg = 200\ g$
7) $0.04\ kg = 40\ g$
8) $0.05\ kg = 50\ g$
9) $0.5\ kg = 500\ g$
10) $3.2\ kg = 3{,}200\ g$
11) $8.2\ kg = 8{,}200\ g$
12) $9.2\ kg = 9{,}200\ g$
13) $35\ kg = 35{,}000\ g$
14) $87\ kg = 87{,}000\ g$
15) $45\ kg = 45{,}000\ g$

16) $15\ kg = 15{,}000\ g$
17) $0.32\ kg = 320\ g$
18) $81\ kg = 81{,}000\ g$
19) $10{,}000\ g = 10\ kg$
20) $20{,}000\ g = 20\ kg$
21) $3{,}000\ g = 3\ kg$
22) $100{,}000\ g = 100\ kg$
23) $150{,}000\ g = 150\ kg$
24) $120{,}000\ g = 120\ kg$
25) $200{,}000\ g = 200\ kg$
26) $30{,}000\ g = 30\ kg$
27) $800{,}000\ g = 800\ kg$
28) $20{,}000\ g = 20\ kg$
29) $40{,}000\ g = 40\ kg$
30) $500{,}000\ g = 500\ kg$

Chapter 10: Time

Topics that you'll practice in this chapter:

- ✓ Read Clocks
- ✓ Telling Time
- ✓ Digital Clock
- ✓ Measurement – Time

Read Clocks

✎ *Write the time below each clock.*

1)

2) _____

3) _____

4)

5) _____

6) _____

✎ *How much time has passed?*

7) From 1:15 AM to 4:35 AM: _____ hours and _____ minutes.

8) From 1:25 AM to 4:05 AM: _____ hours and _____ minutes.

9) It's 8:30 P.M. What time was 5 hours ago?

_____ O'clock

Digital Clock

✎ *What time is it? Write the time in words in front of each.*

1) 2 : 30   ——————————————

2) 3 : 15   ——————————————

3) 5 : 45   ——————————————

4) 9 : 20   ——————————————

5) 10 : 5   ——————————————

6) 12 : 50   ——————————————

7) 10 : 25   ——————————————

8) 3 : 23   ——————————————

9) 11 : 57   ——————————————

10) 2 : 12   ——————————————

11) 1 : 02   ——————————————

12) 8 : 35   ——————————————

Measurement – Time

✎ *How much time has passed?*

1) 1:15 AM to 4:35 AM: _____ hours and _____ minutes.

2) 2:35 AM to 5:10 AM: _____ hours and _____ minutes.

3) 6:00 AM. to 7:25 AM. = _____ hour(s) and _____ minutes.

4) 6:15 PM to 7:30 PM. = _____ hour(s) and _____ minutes

5) 5:15 A.M. to 5:45 A.M. = _____ minutes

6) 4:05 A.M. to 4:30 A.M. = _____ minutes

7) There are _____ second in 15 minutes.

8) There are _____ second in 11 minutes.

9) There are _____ second in 27 minutes.

10) There are _____ minutes in 10 hours.

11) There are _____ minutes in 20 hours.

12) There are _____ minutes in 12 hours.

Answers of Worksheets – Chapter 10

Read clocks

1) 1
2) 4 : 45
3) 8
4) 3 : 30
5) 10 : 15
6) 8 : 35
7) 3 hours and 20 minutes
8) 2 hours and 40 minutes
9) 3 : 30 PM

Digital Clock

1) It's two thirty.
2) It's three Fifteen.
3) It's five forty–five.
4) It's nine twenty.
5) It's ten five.
6) It's Twelve Fifty.
7) It's ten Twenty–five.
8) It's three Twenty–three.
9) It's Eleven fifty seven.
10) It's two Twelve.
11) It's one two.
12) It's eight thirty five.

Measurement – Time

1) 3 : 20
2) 2 : 35
3) 1 : 25
4) 1 : 15
5) 30 minutes
6) 25 minutes
7) 900
8) 660
9) 1,620
10) 600
11) 1,200
12) 720

STAAR Math Exercise Book for Grade 5

Chapter 11: Geometric

Topics that you'll practice in this chapter:

- ✓ Identifying Angles: Acute, Right, Obtuse, and Straight Angles
- ✓ Polygon Names
- ✓ Classify Triangles
- ✓ Parallel Sides in Quadrilaterals
- ✓ Identify Rectangles
- ✓ Perimeter: Find the Missing Side Lengths
- ✓ Perimeter and Area of Squares
- ✓ Perimeter and Area of rectangles
- ✓ Find the Area or Missing Side Length of a Rectangle
- ✓ Area and Perimeter: Word Problems
- ✓ Area of Squares and Rectangles
- ✓ Volume of Cubes and Rectangle Prisms

Identifying Angles: Acute, Right, Obtuse, and Straight Angles

✎ *Write the name of the angles.*

1)

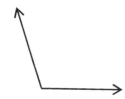

2)

3)

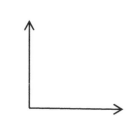

4)

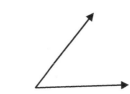

5)

6)

7)

8)

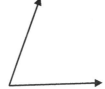

Polygon Names

✍ *Write name of polygons.*

1)

2)

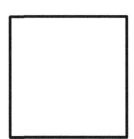

3)

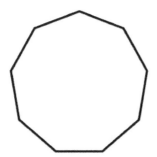

4)

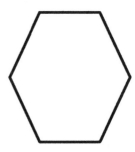

5)

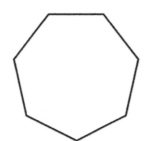

6)

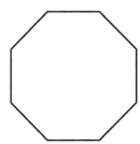

7)

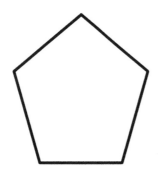

8)

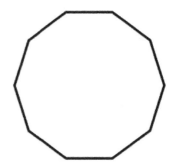

Triangles

 *Classify the triangles by their sides and angles.*

1)

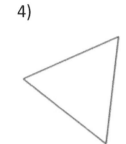

2)

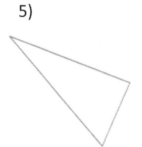

3)

4)

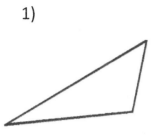

5)

6)

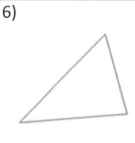

 *Find the measure of the unknown angle in each triangle.*

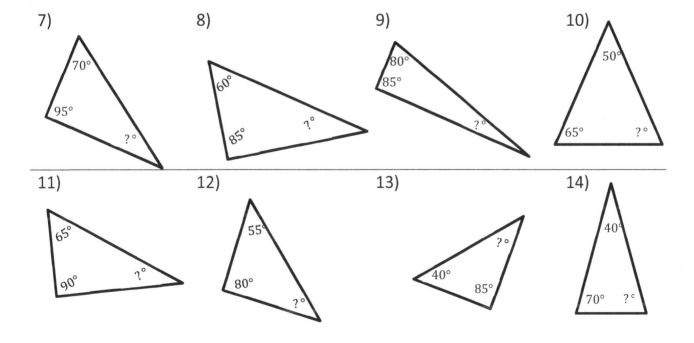

7) 70° 95° ?°

8) 60° 85° ?°

9) 80° 85° ?°

10) 50° 65° ?°

11) 65° 90° ?°

12) 55° 80° ?°

13) ?° 40° 85°

14) 40° 70° ?°

Quadrilaterals and Rectangles

 Write the name of quadrilaterals.

1)

2)

3)

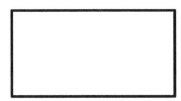

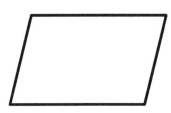

4)

5)

6)

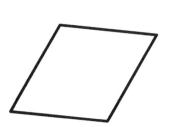

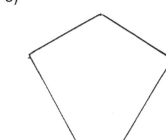

 Solve.

7) A rectangle has _____ sides and _____ angles.

8) Draw a rectangle that is 6 centimeters long and 3 centimeters wide. What is the perimeter?

9) Draw a rectangle 5 cm long and 2 cm wide.

10) Draw a rectangle whose length is 4 cm and whose width is 2 cm. What is the perimeter of the rectangle?

11) What is the perimeter of the rectangle?

8

4

Perimeter: Find the Missing Side Lengths

✎ *Find the missing side of each shape.*

1) perimeter = 44

2) perimeter = 28

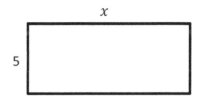

3) perimeter = 30

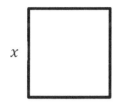

4) perimeter = 16

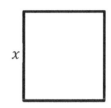

5) perimeter = 60

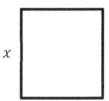

6) perimeter = 22

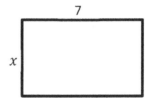

7) perimeter = 30

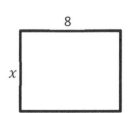

8) perimeter = 36

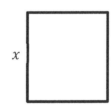

9) perimeter = 50

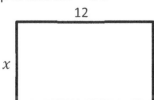

10) perimeter = 48

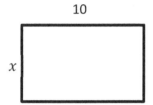

Perimeter and Area of Squares

 *Find perimeter and area of squares.*

1) A: _____ , P: _____

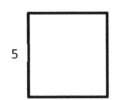
5

2) A: _____ , P: _____

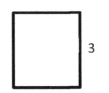
3

3) A: _____ , P: _____

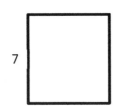
7

4) A: _____ , P: _____

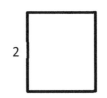
2

5) A: _____ , P: _____

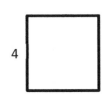
4

6) A: _____ , P: _____

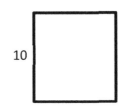
10

7) A: _____ , P: _____

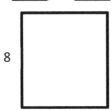
8

8) A: _____ , P: _____

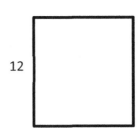
12

9) A: _____ , P: _____

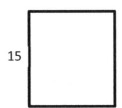
15

10) A: _____ , P: _____

18

Perimeter and Area of rectangles

 *Find perimeter and area of rectangles.*

1) A: ___ , P: ___

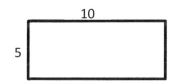

10
5

2) A: ___ , P: ___

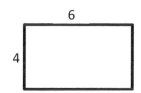

6
4

3) A: ___ , P: ___

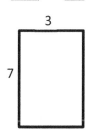

3
7

4) A: ___ , P: ___

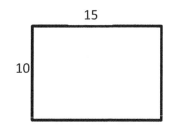

15
10

5) A: ___ , P: ___

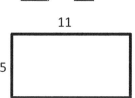

11
5

6) A: ___ , P: ___

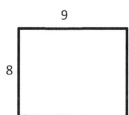

9
8

7) A: ___ , P: ___

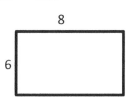

8
6

8) A: ___ , P: ___

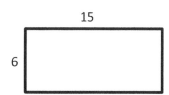

15
6

9) A: ___ , P: ___

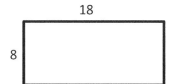

18
8

10) A: ___ , P: ___

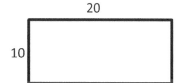

20
10

Find the Area or Missing Side Length of a Rectangle

✍ *Find area or missing side length of rectangles.*

1) Area = ?

14

5

2) Area = 48, x = ?

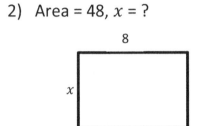

8

x

3) Area = 40, x = ?

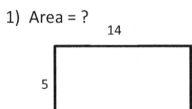

4

x

4) Area = ?

12

8

5) Area = ?

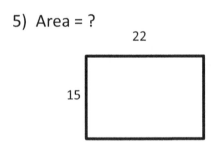

22

15

6) Area = 600, x = ?

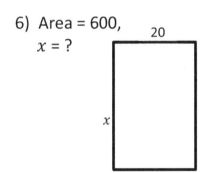

20

x

7) Area = 384, x = ?

32

x

8) Area = 525, x = ?

x

21

9) Area = 450, x = ?

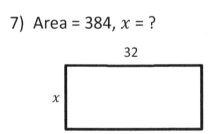

30

x

10) Area = 990, x = ?

55

x

Area and Perimeter

✍ *Find the area of each.*

1)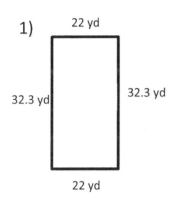

22 yd

32.3 yd   32.3 yd

22 yd

2)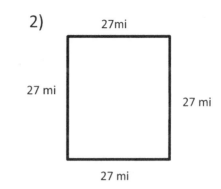

27mi

27 mi   27 mi

27 mi

✍ *Solve.*

3) The area of a rectangle is 72 square meters. The width is 8 meters. What is the length of the rectangle?

4) A square has an area of 36 square feet. What is the perimeter of the square?

5) Ava built a rectangular vegetable garden that is 6 feet long and has an area of 54 square feet. What is the perimeter of Ava's vegetable garden?

6) A square has a perimeter of 64 millimeters. What is the area of the square?

7) The perimeter of David's square backyard is 44 meters. What is the area of David's backyard?

8) The area of a rectangle is 40 square inches. The length is 8 inches. What is the perimeter of the rectangle?

Volume of Cubes

 Find the volume of each cube.

1)

2)
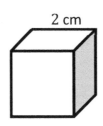
2 cm

3)
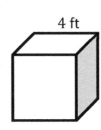
4 ft

4)
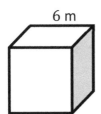
6 m

5)
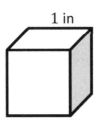
1 in

6)
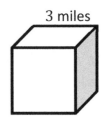
3 miles

7)
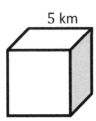
5 km

8)
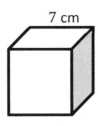
7 cm

9)
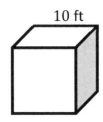
10 ft

10)
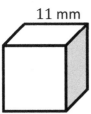
11 mm

11)
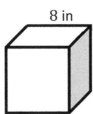
8 in

12)
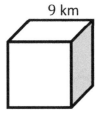
9 km

Answers of Worksheets – Chapter 11

Identifying Angles: Acute, Right, Obtuse, and Straight Angles

1) Obtuse	4) Acute	7) Obtuse
2) Acute	5) Straight	8) Acute
3) Right	6) Obtuse	

Polygon Names

1) Triangle	4) Hexagon	7) Nonagon
2) Quadrilateral	5) Heptagon	8) Decagon
3) Pentagon	6) Octagon	

Triangles

1) Scalene, obtuse	6) Scalene, acute	11) 25°
2) Isosceles, right	7) 15°	12) 45°
3) Scalene, right	8) 35°	13) 55°
4) Equilateral, acute	9) 15°	14) 70°
5) Scalene, acute	10) 65°	

Quadrilaterals and Rectangles

1) Square	5) Trapezoid	9) Use a rule to draw the rectangle
2) Rectangle	6) Kike	10) 12
3) Parallelogram	7) 4 — 4	11) 24
4) Rhombus	8) 18	

Perimeter: Find the Missing Side Lengths

1) 11	5) 15	9) 13
2) 9	6) 4	10) 18
3) 5	7) 7	
4) 4	8) 9	

Perimeter and Area of Squares

1) $A: 25, P: 20$	5) $A: 16, P: 16$	9) $A: 225, P: 60$
2) $A: 9, P: 12$	6) $A: 100, P: 40$	10) $A: 324, P: 72$
3) $A: 49, P: 28$	7) $A: 64, P: 32$	
4) $A: 4, P: 8$	8) $A: 144, P: 48$	

Perimeter and Area of rectangles

1) $A: 50, P: 30$
2) $A: 24, P: 20$
3) $A: 21, P: 20$
4) $A: 150, P: 50$

5) $A: 55, P: 32$
6) $A: 72, P: 34$
7) $A: 48, P: 28$
8) $A: 90, P: 42$

9) $A: 144, P: 52$
10) $A: 200, P: 60$

Find the Area or Missing Side Length of a Rectangle

1) 70
2) 6
3) 10
4) 96
5) 330

6) 30
7) 12
8) 25
9) 15
10) 18

Area and Perimeter

1) $710.6 \, yd^2$
2) $729 \, mi^2$
3) 9

4) 24
5) 30
6) 256
7) 121
8) 26

Volume of Cubes

1) $50.24 \, in^2$
2) $113.04 \, cm^2$
3) $12.56 \, ft^2$
4) $314 \, m^2$
5) $28.26 \, cm^2$
6) $200.96 \, miles^2$

7) $12.56 \, in^2$
8) $3.14 \, ft^2$
9) $50.24 \, m^2$
10) $78.5 \, cm^2$
11) $113.04 \, miles^2$
12) $19.63 \, ft^2$

Chapter 12: Three-Dimensional Figures

Topics that you'll practice in this chapter:

- ✓ Identify Three–Dimensional Figures
- ✓ Count Vertices, Edges, and Faces
- ✓ Identify Faces of Three–Dimensional Figures

Identify Three–Dimensional Figures

✍ *Write the name of each shape.*

1)

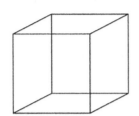

2)

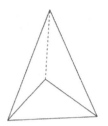

3)

4)

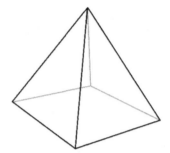

5)

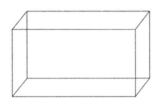

6)

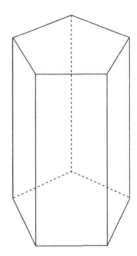

7)

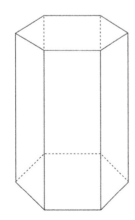

Count Vertices, Edges, and Faces

	Number of edges	Number of faces	Number of vertices
1) 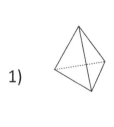	_____	_____	_____
2) 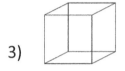	_____	_____	_____
3)	_____	_____	_____
4)	_____	_____	_____
5) 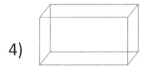	_____	_____	_____
6) 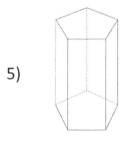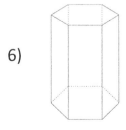	_____	_____	_____

Identify Faces of Three–Dimensional Figures

✎ *Write the number of faces.*

1)

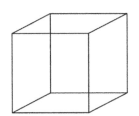

2)

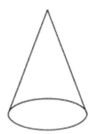

3)

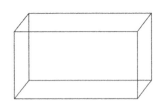

4)

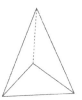

5)

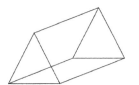

6)

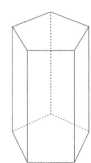

7)

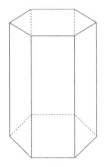

8)

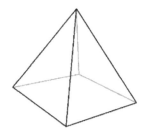

Answers of Worksheets – Chapter 12

Identify Three–Dimensional Figures

1) Cube
2) Triangular pyramid
3) Triangular prism
4) Square pyramid

5) Rectangular prism
6) Pentagonal prism
7) Hexagonal prism

Count Vertices, Edges, and Faces

	Number of edges	Number of faces	Number of vertices
1)	6	4	4
2)	8	5	5
3)	12	6	8
4)	12	6	8

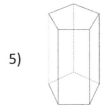

5)   15   7   10

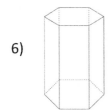

6)   18   8   12

Identify Faces of Three–Dimensional Figures

1) 6
2) 2
3) 5
4) 4
5) 6
6) 7
7) 8
8) 5

Chapter 13: Symmetry and Lines

Topics that you'll practice in this chapter:

- ✓ Line Segments
- ✓ Identify Lines of Symmetry
- ✓ Count Lines of Symmetry
- ✓ Parallel, Perpendicular and Intersecting Lines

Line Segments

✍ *Write each as a line, ray or line segment.*

1)

2)

3)

4)

5)

6)

7)

8)

Identify Lines of Symmetry

✍ *Tell whether the line on each shape is a line of symmetry.*

1)

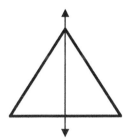

2)

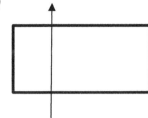

3)

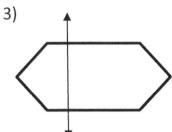

4)

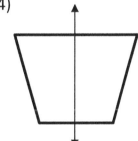

5)

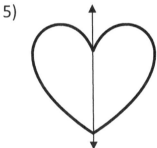

6)

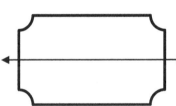

7)

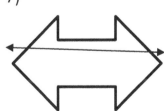

8)

Count Lines of Symmetry

✎ *Draw lines of symmetry on each shape. Count and write the lines of symmetry you see.*

1)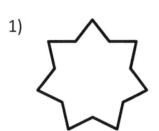

2)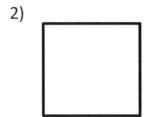

3)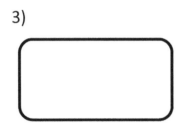

4)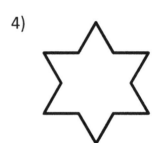

5)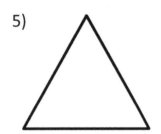

6)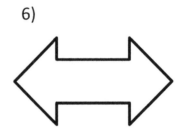

7)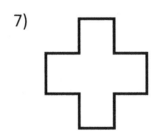

8)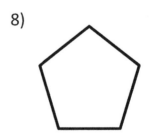

Parallel, Perpendicular and Intersecting Lines

✍ *State whether the given pair of lines are parallel, perpendicular, or intersecting.*

1)

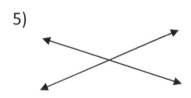

2)

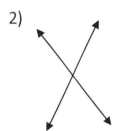

3)

4)

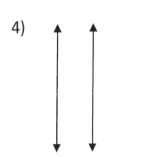

5)

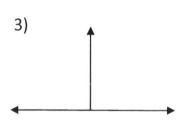

6)

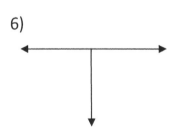

7)

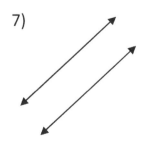

8)

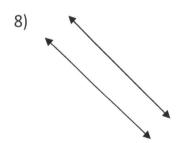

Answers of Worksheets – Chapter 13

Line Segments

1) Line segment
2) Ray
3) Line
4) Line segment

5) Ray
6) Line
7) Line
8) Line segment

Identify lines of symmetry

1) yes
2) no
3) no

4) yes
5) yes
6) yes

7) no
8) yes

Count lines of symmetry

1)

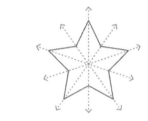

2)

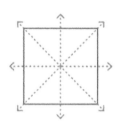

3)

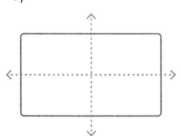

4)

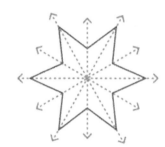

5)

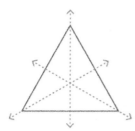

6)

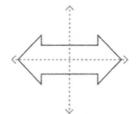

7)

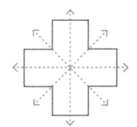

8)

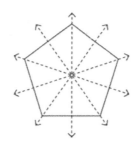

Parallel, Perpendicular and Intersecting Lines

1) Parallel
2) Intersection
3) Perpendicular
4) Parallel

5) Intersection
6) Perpendicular
7) Parallel
8) Parallel

Chapter 14: Fractions

Topics that you'll practice in this chapter:

- ✓ Fractions
- ✓ Add Fractions with Like Denominators
- ✓ Subtract Fractions with Like Denominators
- ✓ Add and Subtract Fractions with Like Denominators
- ✓ Compare Sums and Differences of Fractions with Like Denominators
- ✓ Add 3 or More Fractions with Like Denominators
- ✓ Simplifying Fractions
- ✓ Add Fractions with Unlike Denominators
- ✓ Subtract Fractions with Unlike Denominators
- ✓ Add Fractions with Denominators of 10 and 100
- ✓ Add and Subtract Fractions with Denominators of 10, 100, and 1000

Add Fractions with Like Denominators

✎ *Add fractions.*

1) $\dfrac{2}{3} + \dfrac{1}{3} =$

2) $\dfrac{3}{5} + \dfrac{2}{5} =$

3) $\dfrac{5}{8} + \dfrac{4}{8} =$

4) $\dfrac{3}{4} + \dfrac{3}{4} =$

5) $\dfrac{4}{10} + \dfrac{3}{10} =$

6) $\dfrac{3}{7} + \dfrac{2}{7} =$

7) $\dfrac{4}{5} + \dfrac{4}{5} =$

8) $\dfrac{5}{14} + \dfrac{7}{14} =$

9) $\dfrac{5}{18} + \dfrac{11}{18} =$

10) $\dfrac{3}{12} + \dfrac{5}{12} =$

11) $\dfrac{5}{13} + \dfrac{5}{13} =$

12) $\dfrac{8}{25} + \dfrac{12}{25} =$

13) $\dfrac{9}{15} + \dfrac{6}{15} =$

14) $\dfrac{4}{20} + \dfrac{5}{20} =$

15) $\dfrac{9}{17} + \dfrac{3}{17} =$

16) $\dfrac{18}{32} + \dfrac{15}{32} =$

17) $\dfrac{12}{28} + \dfrac{10}{28} =$

18) $\dfrac{4}{20} + \dfrac{8}{20} =$

19) $\dfrac{24}{45} + \dfrac{11}{45} =$

20) $\dfrac{8}{36} + \dfrac{18}{36} =$

21) $\dfrac{19}{30} + \dfrac{12}{30} =$

22) $\dfrac{23}{42} + \dfrac{10}{42} =$

Subtract Fractions with Like Denominators

✍ *Subtract fractions.*

1) $\dfrac{4}{5} - \dfrac{2}{5} =$

2) $\dfrac{2}{3} - \dfrac{1}{3} =$

3) $\dfrac{7}{9} - \dfrac{4}{9} =$

4) $\dfrac{5}{6} - \dfrac{3}{6} =$

5) $\dfrac{4}{10} - \dfrac{3}{10} =$

6) $\dfrac{5}{7} - \dfrac{3}{7} =$

7) $\dfrac{7}{8} - \dfrac{5}{8} =$

8) $\dfrac{11}{13} - \dfrac{9}{13} =$

9) $\dfrac{8}{10} - \dfrac{5}{10} =$

10) $\dfrac{8}{12} - \dfrac{7}{12} =$

11) $\dfrac{18}{21} - \dfrac{12}{21} =$

12) $\dfrac{15}{19} - \dfrac{9}{19} =$

13) $\dfrac{9}{25} - \dfrac{6}{25} =$

14) $\dfrac{25}{32} - \dfrac{17}{32} =$

15) $\dfrac{22}{27} - \dfrac{9}{27} =$

16) $\dfrac{27}{30} - \dfrac{15}{30} =$

17) $\dfrac{31}{33} - \dfrac{26}{33} =$

18) $\dfrac{18}{28} - \dfrac{8}{28} =$

19) $\dfrac{35}{40} - \dfrac{15}{40} =$

20) $\dfrac{29}{35} - \dfrac{19}{35} =$

21) $\dfrac{21}{36} - \dfrac{11}{36} =$

22) $\dfrac{18}{27} - \dfrac{13}{27} =$

Add and Subtract Fractions with Like Denominators

✍ *Add fractions.*

1) $\frac{1}{2} + \frac{1}{2} =$

2) $\frac{1}{3} + \frac{2}{3} =$

3) $\frac{3}{6} + \frac{2}{6} =$

4) $\frac{5}{8} + \frac{2}{8} =$

5) $\frac{3}{9} + \frac{5}{9} =$

6) $\frac{4}{10} + \frac{1}{10} =$

7) $\frac{3}{7} + \frac{2}{7} =$

8) $\frac{3}{5} + \frac{2}{5} =$

9) $\frac{1}{12} + \frac{1}{12} =$

10) $\frac{16}{25} + \frac{5}{25} =$

✍ *Subtract fractions.*

11) $\frac{4}{5} - \frac{2}{5} =$

12) $\frac{5}{7} - \frac{3}{7} =$

13) $\frac{3}{4} - \frac{2}{4} =$

14) $\frac{8}{9} - \frac{3}{9} =$

15) $\frac{6}{14} - \frac{3}{14} =$

16) $\frac{4}{15} - \frac{1}{15} =$

17) $\frac{15}{16} - \frac{13}{16} =$

18) $\frac{25}{50} - \frac{20}{50} =$

19) $\frac{10}{21} - \frac{7}{21} =$

20) $\frac{12}{27} - \frac{8}{27} =$

Compare Sums and Differences of Fractions with Like Denominators

✎ *Evaluate and compare. Write < or > or =.*

1) $\frac{1}{2} + \frac{1}{2} \ \square \ \frac{1}{3}$

2) $\frac{1}{4} + \frac{2}{4} \ \square \ 1$

3) $\frac{1}{3} + \frac{1}{3} \ \square \ \frac{2}{3}$

4) $\frac{1}{4} + \frac{2}{4} \ \square \ \frac{1}{4}$

5) $\frac{3}{5} + \frac{2}{5} \ \square \ \frac{4}{5}$

6) $\frac{5}{7} - \frac{3}{7} \ \square \ \frac{6}{7}$

7) $\frac{5}{10} + \frac{2}{10} \ \square \ \frac{5}{10}$

8) $\frac{5}{9} - \frac{3}{9} \ \square \ \frac{7}{9}$

9) $\frac{10}{12} - \frac{5}{12} \ \square \ \frac{3}{12}$

10) $\frac{3}{8} + \frac{1}{8} \ \square \ \frac{1}{8}$

11) $\frac{10}{15} + \frac{4}{15} \ \square \ \frac{9}{15}$

12) $\frac{15}{18} - \frac{3}{18} \ \square \ \frac{17}{18}$

13) $\frac{17}{21} + \frac{4}{21} \ \square \ \frac{18}{21}$

14) $\frac{14}{16} - \frac{4}{16} \ \square \ \frac{12}{16}$

15) $\frac{27}{32} - \frac{11}{32} \ \square \ \frac{20}{32}$

16) $\frac{25}{30} + \frac{5}{30} \ \square \ \frac{15}{30}$

17) $\frac{25}{27} - \frac{3}{27} \ \square \ \frac{9}{27}$

18) $\frac{42}{45} - \frac{15}{45} \ \square \ \frac{30}{45}$

19) $\frac{28}{36} + \frac{5}{36} \ \square \ \frac{18}{36}$

20) $\frac{18}{42} + \frac{13}{42} \ \square \ \frac{30}{42}$

Add 3 or More Fractions with Like Denominators

✎ *Add fractions.*

1) $\dfrac{1}{3} + \dfrac{1}{3} + \dfrac{1}{3} =$

2) $\dfrac{2}{5} + \dfrac{1}{5} + \dfrac{1}{5} =$

3) $\dfrac{1}{6} + \dfrac{2}{6} + \dfrac{2}{6} =$

4) $\dfrac{4}{7} + \dfrac{2}{7} + \dfrac{1}{7} =$

5) $\dfrac{1}{5} + \dfrac{3}{5} + \dfrac{1}{5} =$

6) $\dfrac{3}{9} + \dfrac{3}{9} + \dfrac{1}{9} =$

7) $\dfrac{1}{4} + \dfrac{1}{4} + \dfrac{1}{4} =$

8) $\dfrac{7}{15} + \dfrac{3}{15} + \dfrac{4}{15} =$

9) $\dfrac{3}{12} + \dfrac{2}{12} + \dfrac{3}{12} =$

10) $\dfrac{4}{10} + \dfrac{2}{10} + \dfrac{1}{10} =$

11) $\dfrac{5}{18} + \dfrac{5}{18} + \dfrac{3}{18} =$

12) $\dfrac{5}{21} + \dfrac{11}{21} + \dfrac{3}{21} =$

13) $\dfrac{8}{20} + \dfrac{4}{20} + \dfrac{3}{20} =$

14) $\dfrac{2}{16} + \dfrac{5}{16} + \dfrac{8}{16} =$

15) $\dfrac{4}{25} + \dfrac{4}{25} + \dfrac{4}{25} =$

16) $\dfrac{12}{30} + \dfrac{7}{30} + \dfrac{5}{30} =$

17) $\dfrac{9}{27} + \dfrac{6}{27} + \dfrac{6}{27} =$

18) $\dfrac{3}{42} + \dfrac{5}{42} + \dfrac{6}{42} =$

19) $\dfrac{11}{32} + \dfrac{8}{32} + \dfrac{6}{32} =$

20) $\dfrac{9}{37} + \dfrac{11}{37} + \dfrac{10}{37} =$

21) $\dfrac{19}{45} + \dfrac{10}{45} + \dfrac{5}{45} =$

22) $\dfrac{22}{50} + \dfrac{12}{50} + \dfrac{11}{50} =$

Simplifying Fractions

✎ Simplify each fraction to its lowest terms.

1) $\frac{9}{18} =$

2) $\frac{8}{10} =$

3) $\frac{6}{8} =$

4) $\frac{5}{20} =$

5) $\frac{18}{24} =$

6) $\frac{6}{9} =$

7) $\frac{12}{15} =$

8) $\frac{4}{16} =$

9) $\frac{18}{36} =$

10) $\frac{6}{42} =$

11) $\frac{13}{39} =$

12) $\frac{21}{28} =$

13) $\frac{63}{77} =$

14) $\frac{36}{40} =$

15) $\frac{21}{63} =$

16) $\frac{30}{84} =$

17) $\frac{50}{125} =$

18) $\frac{72}{108} =$

19) $\frac{49}{112} =$

20) $\frac{240}{320} =$

21) $\frac{120}{150} =$

✎ Solve each problem.

22) Which of the following fractions equal to $\frac{4}{5}$? _____

A. $\frac{64}{75}$ B. $\frac{92}{115}$ C. $\frac{60}{85}$ D. $\frac{160}{220}$

23) Which of the following fractions equal to $\frac{3}{7}$? _____

A. $\frac{63}{147}$ B. $\frac{75}{182}$ C. $\frac{54}{140}$ D. $\frac{39}{98}$

24) Which of the following fractions equal to $\frac{2}{9}$? _____

A. $\frac{84}{386}$ B. $\frac{52}{234}$ C. $\frac{96}{450}$ D. $\frac{112}{522}$

Add and Subtract Fractions with Unlike Denominators

✎ *Find the sum.*

1) $\frac{1}{3} + \frac{2}{3} =$

2) $\frac{1}{2} + \frac{1}{3} =$

3) $\frac{2}{5} + \frac{1}{2} =$

4) $\frac{3}{7} + \frac{2}{3} =$

5) $\frac{3}{4} + \frac{2}{5} =$

6) $\frac{3}{5} + \frac{1}{5} =$

7) $\frac{5}{9} + \frac{1}{2} =$

8) $\frac{3}{5} + \frac{3}{8} =$

9) $\frac{5}{9} + \frac{3}{7} =$

10) $\frac{5}{11} + \frac{1}{4} =$

11) $\frac{3}{7} + \frac{1}{6} =$

12) $\frac{3}{14} + \frac{3}{4} =$

✎ *Find the difference.*

13) $\frac{1}{2} - \frac{1}{3} =$

14) $\frac{4}{5} - \frac{2}{3} =$

15) $\frac{2}{3} - \frac{1}{6} =$

16) $\frac{3}{5} - \frac{1}{2} =$

17) $\frac{8}{9} - \frac{2}{5} =$

18) $\frac{4}{7} - \frac{1}{9} =$

19) $\frac{2}{5} - \frac{1}{4} =$

20) $\frac{5}{8} - \frac{2}{6} =$

21) $\frac{4}{15} - \frac{1}{10} =$

22) $\frac{7}{20} - \frac{1}{5} =$

23) $\frac{3}{18} - \frac{1}{12} =$

24) $\frac{9}{24} - \frac{3}{16} =$

25) $\frac{3}{7} - \frac{2}{5} =$

26) $\frac{5}{9} - \frac{1}{6} =$

27) $\frac{2}{5} - \frac{1}{10} =$

28) $\frac{5}{12} - \frac{2}{9} =$

29) $\frac{2}{13} - \frac{3}{7} =$

30) $\frac{4}{11} - \frac{5}{8} =$

Add Fractions with Denominators of 10 and 100

✍ *Add fractions.*

1) $\dfrac{5}{10} + \dfrac{20}{100} =$

2) $\dfrac{2}{10} + \dfrac{35}{100} =$

3) $\dfrac{25}{100} + \dfrac{6}{10} =$

4) $\dfrac{73}{100} + \dfrac{1}{10} =$

5) $\dfrac{68}{100} + \dfrac{2}{10} =$

6) $\dfrac{4}{10} + \dfrac{40}{100} =$

7) $\dfrac{80}{100} + \dfrac{1}{10} =$

8) $\dfrac{50}{100} + \dfrac{3}{10} =$

9) $\dfrac{59}{100} + \dfrac{3}{10} =$

10) $\dfrac{7}{10} + \dfrac{12}{100} =$

11) $\dfrac{9}{10} + \dfrac{10}{100} =$

12) $\dfrac{40}{100} + \dfrac{3}{10} =$

13) $\dfrac{36}{100} + \dfrac{4}{10} =$

14) $\dfrac{27}{100} + \dfrac{6}{10} =$

15) $\dfrac{55}{100} + \dfrac{3}{10} =$

16) $\dfrac{1}{10} + \dfrac{85}{100} =$

17) $\dfrac{17}{100} + \dfrac{6}{10} =$

18) $\dfrac{26}{100} + \dfrac{7}{10} =$

19) $\dfrac{45}{100} + \dfrac{4}{10} =$

20) $\dfrac{5}{10} + \dfrac{30}{100} =$

21) $\dfrac{56}{100} + \dfrac{2}{10} =$

22) $\dfrac{67}{100} + \dfrac{3}{10} =$

Add and Subtract Fractions with Denominators of 10, 100, and 1000

✎ *Evaluate fractions.*

1) $\dfrac{25}{100} - \dfrac{2}{10} =$

11) $\dfrac{820}{1000} + \dfrac{5}{10} =$

2) $\dfrac{45}{100} - \dfrac{3}{10} =$

12) $\dfrac{67}{100} + \dfrac{240}{1000} =$

3) $\dfrac{8}{10} - \dfrac{30}{100} =$

13) $\dfrac{7}{10} - \dfrac{12}{100} =$

4) $\dfrac{6}{10} + \dfrac{27}{100} =$

14) $\dfrac{75}{100} - \dfrac{5}{10} =$

5) $\dfrac{25}{100} + \dfrac{450}{1000} =$

15) $\dfrac{70}{100} - \dfrac{3}{10} =$

6) $\dfrac{73}{100} - \dfrac{320}{1000} =$

16) $\dfrac{850}{1000} - \dfrac{5}{100} =$

7) $\dfrac{25}{100} + \dfrac{670}{1000} =$

17) $\dfrac{300}{1000} + \dfrac{12}{100} =$

8) $\dfrac{4}{10} + \dfrac{780}{1000} =$

18) $\dfrac{780}{1000} - \dfrac{6}{10} =$

9) $\dfrac{80}{100} - \dfrac{560}{1000} =$

19) $\dfrac{80}{100} - \dfrac{6}{10} =$

10) $\dfrac{78}{100} - \dfrac{6}{10} =$

20) $\dfrac{50}{100} - \dfrac{210}{1000} =$

21) $\dfrac{350}{1000} - \dfrac{3}{10} =$

22) $\dfrac{85}{100} - \dfrac{450}{1000} =$

Answers of Worksheets – Chapter 14

Add Fractions with Like Denominators

1) 1

2) 1

3) $\dfrac{9}{8}$

4) $\dfrac{6}{4}$

5) $\dfrac{7}{10}$

6) $\dfrac{5}{7}$

7) $\dfrac{8}{5}$

8) $\dfrac{12}{14}$

9) $\dfrac{16}{18}$

10) $\dfrac{8}{12}$

11) $\dfrac{10}{13}$

12) $\dfrac{20}{25}$

13) 1

14) $\dfrac{9}{20}$

15) $\dfrac{12}{17}$

16) $\dfrac{33}{32}$

17) $\dfrac{22}{28}$

18) $\dfrac{12}{20}$

19) $\dfrac{35}{45}$

20) $\dfrac{26}{36}$

21) $\dfrac{31}{30}$

22) $\dfrac{33}{42}$

Subtract Fractions with Like Denominators

1) $\dfrac{2}{5}$

2) $\dfrac{1}{3}$

3) $\dfrac{3}{9}$

4) $\dfrac{2}{6}$

5) $\dfrac{1}{10}$

6) $\dfrac{2}{7}$

7) $\dfrac{2}{8}$

8) $\dfrac{2}{13}$

9) $\dfrac{3}{10}$

10) $\dfrac{1}{12}$

11) $\dfrac{6}{21}$

12) $\dfrac{6}{19}$

13) $\dfrac{3}{25}$

14) $\dfrac{1}{4}$

15) $\dfrac{13}{27}$

16) $\dfrac{12}{30}$

17) $\dfrac{5}{33}$

18) $\dfrac{10}{28}$

19) $\dfrac{20}{40}$

20) $\dfrac{2}{7}$

21) $\dfrac{10}{36}$

22) $\dfrac{5}{27}$

Add and Subtract Fractions with Like Denominators

1) 1

2) 1

3) $\frac{5}{6}$

4) $\frac{7}{8}$

5) $\frac{8}{9}$

6) $\frac{5}{10}$

7) $\frac{5}{7}$

8) 1

9) $\frac{2}{12}$

10) $\frac{21}{25}$

11) $\frac{2}{5}$

12) $\frac{2}{7}$

13) $\frac{1}{4}$

14) $\frac{5}{9}$

15) $\frac{3}{14}$

16) $\frac{3}{15}$

17) $\frac{2}{16}$

18) $\frac{5}{50}$

19) $\frac{3}{21}$

20) $\frac{4}{27}$

Compare Sums and Differences of Fractions with Like Denominators

1) $1 > \frac{1}{3}$

2) $\frac{3}{4} < 1$

3) $\frac{2}{3} = \frac{2}{3}$

4) $\frac{3}{4} > \frac{1}{4}$

5) $1 > \frac{4}{5}$

6) $\frac{2}{7} < \frac{6}{7}$

7) $\frac{7}{10} > \frac{5}{10}$

8) $\frac{2}{9} < \frac{7}{9}$

9) $\frac{5}{12} > \frac{3}{12}$

10) $\frac{4}{8} > \frac{1}{8}$

11) $\frac{14}{15} > \frac{9}{15}$

12) $\frac{12}{18} < \frac{17}{18}$

13) $1 > \frac{18}{21}$

14) $\frac{10}{16} < \frac{12}{16}$

15) $\frac{16}{32} < \frac{20}{32}$

16) $1 > \frac{15}{30}$

17) $\frac{22}{27} > \frac{9}{27}$

18) $\frac{27}{45} < \frac{30}{45}$

19) $\frac{33}{36} > \frac{18}{36}$

20) $\frac{31}{42} > \frac{30}{42}$

Add 3 or More Fractions with Like Denominators

1) 1

2) $\frac{4}{5}$

3) $\frac{5}{6}$

4) 1

5) 1

6) $\frac{7}{9}$

7) $\frac{3}{4}$

8) $\frac{14}{15}$

9) $\frac{8}{12}$

10) $\frac{7}{10}$

11) $\frac{13}{18}$

12) $\frac{19}{21}$

13) $\frac{15}{20}$

14) $\dfrac{15}{16}$

15) $\dfrac{12}{25}$

16) $\dfrac{24}{30}$

17) $\dfrac{21}{27}$

18) $\dfrac{14}{42}$

19) $\dfrac{25}{32}$

20) $\dfrac{30}{37}$

21) $\dfrac{34}{45}$

22) $\dfrac{45}{50}$

Simplifying Fractions

1) $\dfrac{1}{2}$

2) $\dfrac{4}{5}$

3) $\dfrac{3}{4}$

4) $\dfrac{1}{4}$

5) $\dfrac{3}{4}$

6) $\dfrac{2}{3}$

7) $\dfrac{4}{5}$

8) $\dfrac{1}{4}$

9) $\dfrac{1}{2}$

10) $\dfrac{1}{7}$

11) $\dfrac{1}{3}$

12) $\dfrac{3}{4}$

13) $\dfrac{9}{11}$

14) $\dfrac{9}{10}$

15) $\dfrac{1}{3}$

16) $\dfrac{5}{14}$

17) $\dfrac{2}{5}$

18) $\dfrac{2}{3}$

19) $\dfrac{7}{16}$

20) $\dfrac{3}{4}$

21) $\dfrac{4}{5}$

22) B

23) A

24) B

Add and Subtract fractions with unlike denominators

1) $\dfrac{3}{3} = 1$

2) $\dfrac{5}{6}$

3) $\dfrac{9}{10}$

4) $\dfrac{23}{21}$

5) $\dfrac{23}{20}$

6) $\dfrac{4}{5}$

7) $\dfrac{19}{18}$

8) $\dfrac{39}{40}$

9) $\dfrac{62}{63}$

10) $\dfrac{31}{44}$

11) $\dfrac{25}{42}$

12) $\dfrac{27}{28}$

13) $\dfrac{1}{6}$

14) $\dfrac{2}{15}$

15) $\dfrac{1}{2}$

16) $\dfrac{1}{10}$

17) $\dfrac{22}{45}$

18) $\dfrac{29}{63}$

19) $\dfrac{3}{20}$

20) $\dfrac{7}{24}$

21) $\dfrac{1}{6}$

22) $\dfrac{3}{20}$

23) $\dfrac{1}{12}$

24) $\dfrac{3}{16}$

25) $\dfrac{1}{35}$

26) $\dfrac{7}{18}$

27) $\dfrac{3}{10}$

28) $\dfrac{7}{36}$

29) $-\dfrac{25}{91}$

30) $-\dfrac{15}{88}$

Add fractions with denominators of 10 and 100

1) $\dfrac{7}{10}$

2) $\dfrac{11}{20}$

3) $\dfrac{17}{20}$

4) $\dfrac{83}{100}$

5) $\dfrac{22}{25}$

6) $\dfrac{4}{5}$

7) $\dfrac{9}{10}$

8) $\dfrac{4}{5}$

9) $\dfrac{89}{100}$

10) $\dfrac{41}{50}$

11) 1

12) $\dfrac{7}{10}$

13) $\dfrac{19}{25}$

14) $\dfrac{87}{100}$

15) $\dfrac{17}{20}$

16) $\dfrac{19}{20}$

17) $\dfrac{77}{100}$

18) $\dfrac{24}{25}$

19) $\dfrac{17}{20}$

20) $\dfrac{4}{5}$

21) $\dfrac{19}{25}$

22) $\dfrac{97}{100}$

Add and subtract fractions with denominators of 10, 100, and 1000

1) $\dfrac{1}{20}$

2) $\dfrac{3}{20}$

3) $\dfrac{50}{100}$

4) $\dfrac{87}{100}$

5) $\dfrac{7}{10}$

6) $\dfrac{41}{100}$

7) $\dfrac{23}{25}$

8) $\dfrac{59}{50}$

9) $\dfrac{6}{25}$

10) $\dfrac{9}{50}$

11) $\dfrac{33}{25}$

12) $\dfrac{91}{100}$

13) $\dfrac{29}{50}$

14) $\dfrac{1}{4}$

15) $\dfrac{2}{5}$

16) $\dfrac{4}{5}$

17) $\dfrac{21}{50}$

18) $\dfrac{9}{50}$

19) $\dfrac{1}{5}$

20) $\dfrac{29}{100}$

21) $\dfrac{1}{20}$

22) $\dfrac{2}{5}$

Chapter 15:
Mixed Numbers

Topics that you'll practice in this chapter:

- ✓ Fractions to Mixed Numbers
- ✓ Mixed Numbers to Fractions
- ✓ Add and Subtract Mixed Numbers
- ✓ Multiplying and Dividing Mixed Numbers

Fractions to Mixed Numbers

✎ *Convert fractions to mixed numbers.*

1) $\dfrac{4}{3} =$

2) $\dfrac{3}{2} =$

3) $\dfrac{5}{3} =$

4) $\dfrac{7}{2} =$

5) $\dfrac{8}{5} =$

6) $\dfrac{7}{3} =$

7) $\dfrac{9}{4} =$

8) $\dfrac{12}{5} =$

9) $\dfrac{13}{9} =$

10) $\dfrac{18}{7} =$

11) $\dfrac{15}{7} =$

12) $\dfrac{19}{6} =$

13) $\dfrac{13}{5} =$

14) $\dfrac{37}{5} =$

15) $\dfrac{21}{6} =$

16) $\dfrac{41}{10} =$

17) $\dfrac{11}{2} =$

18) $\dfrac{56}{10} =$

19) $\dfrac{20}{12} =$

20) $\dfrac{9}{5} =$

21) $\dfrac{19}{5} =$

22) $\dfrac{27}{10} =$

23) $\dfrac{10}{6} =$

24) $\dfrac{17}{8} =$

25) $\dfrac{7}{2} =$

26) $\dfrac{39}{4} =$

27) $\dfrac{72}{10} =$

28) $\dfrac{13}{3} =$

29) $\dfrac{45}{8} =$

30) $\dfrac{27}{5} =$

Mixed Numbers to Fractions

✍ *Convert to fraction.*

1) $2\frac{1}{2} =$

2) $1\frac{2}{3} =$

3) $1\frac{1}{3} =$

4) $2\frac{1}{4} =$

5) $3\frac{2}{5} =$

6) $4\frac{1}{4} =$

7) $5\frac{2}{3} =$

8) $1\frac{2}{7} =$

9) $3\frac{2}{9} =$

10) $1\frac{2}{6} =$

11) $2\frac{2}{3} =$

12) $5\frac{1}{3} =$

13) $6\frac{4}{5} =$

14) $2\frac{3}{4} =$

15) $2\frac{5}{7} =$

16) $3\frac{5}{9} =$

17) $2\frac{9}{10} =$

18) $7\frac{5}{6} =$

19) $6\frac{11}{12} =$

20) $8\frac{9}{20} =$

21) $8\frac{2}{5} =$

22) $5\frac{4}{5} =$

23) $9\frac{1}{6} =$

24) $3\frac{3}{4} =$

25) $5\frac{2}{8} =$

26) $10\frac{2}{3} =$

27) $12\frac{3}{4} =$

28) $14\frac{6}{7} =$

29) $3\frac{7}{11} =$

30) $6\frac{5}{11} =$

31) $7\frac{6}{15} =$

32) $9\frac{11}{21} =$

33) $5\frac{15}{27} =$

Add and Subtract Mixed Numbers

✎ *Find the sum.*

1) $2\frac{1}{2} + 1\frac{1}{3} =$

2) $6\frac{1}{2} + 3\frac{1}{2} =$

3) $2\frac{3}{8} + 3\frac{1}{8} =$

4) $4\frac{1}{2} + 1\frac{1}{4} =$

5) $1\frac{3}{7} + 1\frac{5}{14} =$

6) $6\frac{5}{12} + 3\frac{3}{4} =$

7) $5\frac{1}{2} + 8\frac{3}{4} =$

8) $3\frac{7}{8} + 3\frac{1}{3} =$

9) $3\frac{3}{9} + 7\frac{6}{11} =$

10) $7\frac{5}{12} + 4\frac{3}{10} =$

✎ *Find the difference.*

11) $3\frac{1}{3} - 1\frac{1}{3} =$

12) $4\frac{1}{2} - 3\frac{1}{2} =$

13) $5\frac{1}{2} - 2\frac{1}{4} =$

14) $6\frac{1}{6} - 5\frac{1}{3} =$

15) $8\frac{1}{2} - 1\frac{1}{10} =$

16) $9\frac{1}{2} - 2\frac{1}{4} =$

17) $9\frac{1}{5} - 5\frac{1}{6} =$

18) $14\frac{3}{10} - 13\frac{1}{3} =$

19) $19\frac{2}{3} - 11\frac{5}{8} =$

20) $20\frac{3}{4} - 14\frac{2}{3} =$

21) $2\frac{1}{2} - 1\frac{1}{5} =$

22) $3\frac{1}{6} - 1\frac{1}{10} =$

23) $16\frac{2}{7} - 11\frac{2}{3} =$

24) $15\frac{1}{7} - 10\frac{1}{8} =$

25) $12\frac{3}{4} - 7\frac{1}{3} =$

26) $15\frac{2}{5} - 5\frac{2}{3} =$

Multiplying and Dividing Mixed Numbers

✎ *Find the product.*

1) $4\frac{1}{3} \times 2\frac{1}{5} =$

2) $3\frac{1}{2} \times 3\frac{1}{4} =$

3) $5\frac{2}{5} \times 2\frac{1}{3} =$

4) $2\frac{1}{2} \times 1\frac{2}{9} =$

5) $3\frac{4}{7} \times 2\frac{3}{5} =$

6) $7\frac{2}{3} \times 2\frac{2}{3} =$

7) $9\frac{8}{9} \times 8\frac{3}{4} =$

8) $2\frac{4}{7} \times 5\frac{2}{9} =$

9) $5\frac{2}{5} \times 2\frac{3}{5} =$

10) $3\frac{5}{7} \times 3\frac{5}{6} =$

✎ *Find the quotient.*

11) $1\frac{2}{3} \div 3\frac{1}{3} =$

12) $2\frac{1}{4} \div 1\frac{1}{2} =$

13) $10\frac{1}{2} \div 1\frac{2}{3} =$

14) $3\frac{1}{6} \div 4\frac{2}{3} =$

15) $4\frac{1}{8} \div 2\frac{1}{2} =$

16) $2\frac{1}{10} \div 2\frac{3}{5} =$

17) $1\frac{4}{11} \div 1\frac{1}{4} =$

18) $9\frac{1}{2} \div 9\frac{2}{3} =$

19) $8\frac{3}{4} \div 2\frac{2}{5} =$

20) $12\frac{1}{2} \div 9\frac{1}{3} =$

21) $2\frac{1}{8} \div 1\frac{1}{2} =$

22) $1\frac{1}{10} \div 1\frac{3}{5} =$

23) $5\frac{2}{5} \div 1\frac{3}{4} =$

24) $5\frac{1}{2} \div 2\frac{2}{3} =$

25) $3\frac{3}{4} \div 1\frac{1}{5} =$

26) $3\frac{1}{2} \div 1\frac{1}{3} =$

Answers of Worksheets – Chapter 15

Fractions to Mixed Numbers

1) $1\frac{1}{3}$

2) $1\frac{1}{2}$

3) $1\frac{2}{3}$

4) $3\frac{1}{2}$

5) $1\frac{3}{5}$

6) $2\frac{1}{3}$

7) $2\frac{1}{4}$

8) $2\frac{2}{5}$

9) $1\frac{4}{9}$

10) $2\frac{4}{7}$

11) $2\frac{1}{7}$

12) $3\frac{1}{6}$

13) $2\frac{3}{5}$

14) $7\frac{2}{5}$

15) $3\frac{1}{2}$

16) $4\frac{1}{10}$

17) $5\frac{1}{2}$

18) $5\frac{3}{5}$

19) $1\frac{2}{3}$

20) $1\frac{4}{5}$

21) $3\frac{4}{5}$

22) $2\frac{7}{10}$

23) $1\frac{2}{3}$

24) $2\frac{1}{8}$

25) $3\frac{1}{2}$

26) $9\frac{3}{4}$

27) $7\frac{1}{5}$

28) $4\frac{1}{3}$

29) $5\frac{5}{8}$

30) $5\frac{2}{5}$

Mixed Numbers to Fractions

1) $\frac{5}{2}$

2) $\frac{5}{3}$

3) $\frac{4}{3}$

4) $\frac{9}{4}$

5) $\frac{17}{5}$

6) $\frac{17}{4}$

7) $\frac{17}{3}$

8) $\frac{9}{7}$

9) $\frac{29}{9}$

10) $\frac{4}{3}$

11) $\frac{8}{3}$

12) $\frac{16}{3}$

13) $\frac{34}{5}$

14) $\frac{11}{4}$

15) $\frac{19}{7}$

16) $\frac{32}{9}$

17) $\frac{29}{10}$

18) $\frac{47}{6}$

19) $\frac{83}{12}$

20) $\frac{169}{20}$

21) $\frac{42}{5}$

22) $\frac{29}{5}$

23) $\frac{55}{6}$

24) $\frac{21}{4}$

25) $\frac{15}{4}$

26) $\frac{32}{3}$

27) $\frac{51}{4}$

28) $\frac{104}{7}$

29) $\frac{40}{11}$

30) $\frac{71}{11}$

31) $\frac{37}{5}$

32) $\frac{200}{21}$

33) $\frac{50}{9}$

Add and Subtract Mixed Numbers with Like Denominators

1) $3\frac{5}{6}$

2) 10

3) $5\frac{1}{2}$

4) $5\frac{3}{4}$

5) $2\frac{11}{14}$

6) $10\frac{1}{6}$

7) $14\frac{1}{4}$

8) $7\frac{5}{24}$

9) $10\frac{29}{33}$

10) $11\frac{43}{60}$

11) 2

12) 1

13) $3\frac{1}{4}$

14) $\frac{5}{6}$

15) $7\frac{2}{5}$

16) $7\frac{1}{4}$

17) $4\frac{1}{30}$

18) $\frac{29}{30}$

19) $8\frac{1}{24}$

20) $6\frac{1}{12}$

21) $\frac{13}{10}$

22) $2\frac{1}{15}$

23) $4\frac{13}{21}$

24) $5\frac{1}{56}$

25) $5\frac{5}{12}$

26) $9\frac{11}{15}$

Adding and Subtracting Mixed Numbers

1) $3\frac{5}{6}$

2) 10

3) $5\frac{1}{2}$

4) $5\frac{3}{4}$

5) $2\frac{11}{14}$

6) $10\frac{1}{6}$

7) $14\frac{1}{4}$

8) $7\frac{5}{24}$

9) $10\frac{29}{33}$

10) $11\frac{43}{60}$

11) 2

12) 1

13) $3\frac{1}{4}$

14) $\frac{5}{6}$

15) $7\frac{2}{5}$

16) $7\frac{1}{4}$

17) $4\frac{1}{30}$

18) $\frac{29}{30}$

19) $8\frac{1}{24}$

20) $6\frac{1}{12}$

21) $\frac{13}{10}$

22) $2\frac{1}{15}$

23) $4\frac{13}{21}$

24) $5\frac{1}{56}$

25) $5\frac{5}{12}$

26) $9\frac{11}{15}$

Chapter 16:
Proportions, Ratios, and Percent

Topics that you'll practice in this chapter:

✓ Simplifying Ratios

✓ Proportional Ratios

✓ Similarity and Ratios

✓ Ratio and Rates Word Problems

Simplifying Ratios

✎ *Reduce each ratio.*

1) $12 : 8 = $ ___ : ___

2) $2 : 20 = $ ___ : ___

3) $3 : 36 = $ ___ : ___

4) $8 : 16 = $ ___ : ___

5) $6 : 100 = $ ___ : ___

6) $10 : 60 = $ ___ : ___

7) $21 : 49 = $ ___ : ___

8) $20 : 40 = $ ___ : ___

9) $10 : 50 = $ ___ : ___

10) $14 : 18 = $ ___ : ___

11) $45 : 27 = $ ___ : ___

12) $49 : 21 = $ ___ : ___

13) $100 : 10 = $ ___ : ___

14) $35 : 45 = $ ___ : ___

15) $8 : 20 = $ ___ : ___

16) $25 : 35 = $ ___ : ___

17) $21 : 27 = $ ___ : ___

18) $52 : 82 = $ ___ : ___

19) $12 : 36 = $ ___ : ___

20) $24 : 3 = $ ___ : ___

21) $15 : 30 = $ ___ : ___

22) $14 : 63 = $ ___ : ___

23) $68 : 80 = $ ___ : ___

24) $8 : 80 = $ ___ : ___

✎ *Write each ratio as a fraction in simplest form.*

25) $2 : 4 = $

26) $6 : 20 = $

27) $5 : 35 = $

28) $10 : 55 = $

29) $8 : 24 = $

30) $9 : 42 = $

31) $12 : 48 = $

32) $6 : 40 = $

33) $15 : 36 = $

34) $18 : 82 = $

35) $22 : 26 = $

36) $8 : 36 = $

37) $16 : 128 = $

38) $14 : 77 = $

39) $12 : 180 = $

40) $36 : 108 = $

41) $24 : 42 = $

42) $18 : 120 = $

43) $44 : 82 = $

44) $60 : 240 = $

45) $36 : 180 = $

Proportional Ratios

✎ *Fill in the blanks; solve each proportion.*

1) $3 : 7 = \underline{\quad} : 49$

2) $1 : 2 = 20 : \underline{\quad}$

3) $1 : 5 = \underline{\quad} : 50$

4) $7 : 9 = 14 : \underline{\quad}$

5) $5 : 3 = 45 : \underline{\quad}$

6) $7 : 3 = \underline{\quad} : 18$

7) $10 : 1 = \underline{\quad} : 10$

8) $1 : 3 = \underline{\quad} : 27$

9) $8 : 1 = \underline{\quad} : 8$

10) $9 : 2 = \underline{\quad} : 14$

11) $3 : 12 = 12 : \underline{\quad}$

12) $6 : 4 = 24 : \underline{\quad}$

✎ *State if each pair of ratios form a proportion.*

13) $\frac{3}{10}$ and $\frac{9}{30}$

14) $\frac{1}{2}$ and $\frac{16}{32}$

15) $\frac{5}{6}$ and $\frac{35}{42}$

16) $\frac{3}{7}$ and $\frac{27}{72}$

17) $\frac{2}{5}$ and $\frac{16}{45}$

18) $\frac{4}{9}$ and $\frac{40}{81}$

19) $\frac{6}{11}$ and $\frac{42}{77}$

20) $\frac{1}{6}$ and $\frac{8}{48}$

21) $\frac{6}{17}$ and $\frac{36}{85}$

22) $\frac{2}{7}$ and $\frac{24}{86}$

23) $\frac{12}{19}$ and $\frac{156}{247}$

24) $\frac{13}{21}$ and $\frac{182}{294}$

✎ *Solve each proportion.*

25) $\frac{2}{5} = \frac{14}{x}, x = \underline{\quad}$

26) $\frac{1}{6} = \frac{7}{x}, x = \underline{\quad}$

27) $\frac{3}{5} = \frac{27}{x}, x = \underline{\quad}$

28) $\frac{1}{5} = \frac{x}{80}, x = \underline{\quad}$

29) $\frac{3}{7} = \frac{x}{63}, x = \underline{\quad}$

30) $\frac{1}{4} = \frac{13}{x}, x = \underline{\quad}$

31) $\frac{7}{9} = \frac{56}{x}, x = \underline{\quad}$

32) $\frac{6}{11} = \frac{42}{x}, x = \underline{\quad}$

33) $\frac{4}{7} = \frac{x}{77}, x = \underline{\quad}$

34) $\frac{5}{13} = \frac{x}{143}, x = \underline{\quad}$

35) $\frac{7}{19} = \frac{x}{209}, x = \underline{\quad}$

36) $\frac{3}{13} = \frac{x}{195}, x = \underline{\quad}$

Ratio and Rates Word Problems

✍ *Solve each word problem.*

1) Bob has 12 red cards and 20 green cards. What is the ratio of Bob's red cards to his green cards? _____

2) In a party, 10 soft drinks are required for every 12 guests. If there are 252 guests, how many soft drinks is required? _____

3) In Jack's class, 18 of the students are tall and 10 are short. In Michael's class 54 students are tall and 30 students are short. Which class has a higher ratio of tall to short students? _____

4) The price of 3 apples at the Quick Market is $1.44. The price of 5 of the same apples at Walmart is $2.50. Which place is the better buy? _____

5) The bakers at a Bakery can make 160 bagels in 4 hours. How many bagels can they bake in 16 hours? What is that rate per hour? _____

6) You can buy 5 cans of green beans at a supermarket for $3.40. How much does it cost to buy 35 cans of green beans? _____

7) The ratio of boys to girls in a class is 2:3. If there are 18 boys in the class, how many girls are in that class? _____

8) The ratio of red marbles to blue marbles in a bag is 3:4. If there are 42 marbles in the bag, how many of the marbles are red? _____

Answers of Worksheets – Chapter 16

Simplifying Ratios

1) 3 : 2
2) 1 : 10
3) 1 : 12
4) 1 : 2
5) 3 : 50
6) 1 : 6
7) 3 : 7
8) 1 : 2
9) 1 : 5
10) 7 : 9
11) 5 : 3
12) 7 : 3
13) 10 : 1
14) 7 : 9
15) 2 : 5
16) 5 : 7
17) 7 : 9
18) 26 : 41

19) 1 : 3
20) 8 : 1
21) 1 : 2
22) 2 : 9
23) 17 : 20
24) 1 : 10
25) $\frac{1}{2}$
26) $\frac{3}{10}$
27) $\frac{1}{7}$
28) $\frac{2}{11}$
29) $\frac{1}{3}$
30) $\frac{3}{14}$
31) $\frac{1}{4}$
32) $\frac{3}{20}$

33) $\frac{5}{12}$
34) $\frac{9}{41}$
35) $\frac{11}{13}$
36) $\frac{2}{9}$
37) $\frac{1}{8}$
38) $\frac{2}{11}$
39) $\frac{1}{15}$
40) $\frac{1}{3}$
41) $\frac{4}{7}$
42) $\frac{3}{20}$
43) $\frac{22}{41}$
44) $\frac{1}{4}$
45) $\frac{1}{5}$

Proportional Ratios

1) 21
2) 40
3) 10
4) 18
5) 27
6) 42
7) 100
8) 9
9) 64
10) 63
11) 48
12) 16

13) Yes
14) Yes
15) Yes
16) No
17) No
18) No
19) Yes
20) Yes
21) No
22) No
23) Yes
24) Yes

25) 35
26) 42
27) 45
28) 16
29) 27
30) 52
31) 72
32) 77
33) 44
34) 55
35) 77
36) 45

Ratio and Rates Word Problems

1) $3:5$

2) 210

3) The ratio for both classes is 9 to 5.

4) Quick Market is a better buy.

5) 640, the rate is 40 per hour.

6) $23.80

7) 27

8) 18

Chapter 17: Decimal

Topics that you'll practice in this chapter:

- ✓ Decimal Place Value
- ✓ Ordering and Comparing Decimals
- ✓ Decimal Addition
- ✓ Decimal Subtraction

Decimal Place Value

✎ *What place is the selected digit?*

1) 1,122.25

2) 2,321.32

3) 4,258.91

4) 6,372.67

5) 7,131.98

6) 5,442.73

7) 1,841.89

8) 5,995.76

9) 8,982.55

10) 1,249.21

11) 4,316.50

12) 9,191.99

13) 9,112.51

14) 8,435.27

15) 1,662.24

16) 1,148.44

17) 9,989.69

18) 3,155.91

✎ *What is the value of the selected digit?*

19) 3,122.31

20) 1,318.66

21) 6,352.25

22) 3,739.16

23) 4,936.78

24) 7,625.86

25) 9,313.45

26) 2,168.82

27) 8,451.76

28) 2,153.23

Ordering and Comparing Decimals

✍ **Write the correct comparison symbol (>, < or =).**

1) 0.50 ☐ 0.050

2) 0.025 ☐ 0.25

3) 2.060 ☐ 2.07

4) 1.75 ☐ 1.07

5) 4.04 ☐ 0.440

6) 3.05 ☐ 3.5

7) 5.05 ☐ 5.050

8) 1.02 ☐ 1.1

9) 2.45 ☐ 2.125

10) 0.932 ☐ 0.0932

11) 3.15 ☐ 3.150

12) 0.718 ☐ 0.89

13) 7.060 ☐ 7.60

14) 3.59 ☐ 3.129

15) 4.33 ☐ 4.319

16) 2.25 ☐ 2.250

17) 1.95 ☐ 1.095

18) 8.051 ☐ 8.50

✍ **Order each set of integers from least to greatest.**

19) 0.4, 0.54, 0.23, 0.87, 0.36   ___, ___, ___, ___, ___, ___

20) 1.2, 2.4, 1.97, 3.65, 1.80   ___, ___, ___, ___, ___, ___

21) 2.3, 1.2, 1.9, 0.67, 0.34   ___, ___, ___, ___, ___, ___

22) 1.7, 1.2, 3.2, 4.2, 1.34, 3.55   ___, ___, ___, ___, ___, ___

Adding and Subtracting Decimals

✍ *Add and subtract decimals.*

1)
$$
\begin{array}{r}
31.13 \\
-\ 11.45 \\
\hline
\end{array}
$$

4)
$$
\begin{array}{r}
56.67 \\
-\ 44.39 \\
\hline
\end{array}
$$

7)
$$
\begin{array}{r}
66.24 \\
-\ 23.11 \\
\hline
\end{array}
$$

2)
$$
\begin{array}{r}
35.25 \\
+\ 24.47 \\
\hline
\end{array}
$$

5)
$$
\begin{array}{r}
71.47 \\
+\ 16.25 \\
\hline
\end{array}
$$

8)
$$
\begin{array}{r}
39.75 \\
+\ 12.85 \\
\hline
\end{array}
$$

3)
$$
\begin{array}{r}
73.50 \\
+\ 22.78 \\
\hline
\end{array}
$$

6)
$$
\begin{array}{r}
68.99 \\
-\ 53.61 \\
\hline
\end{array}
$$

9)
$$
\begin{array}{r}
229.25 \\
-\ 84.67 \\
\hline
\end{array}
$$

✍ *Find the missing number.*

10) ___ + 2.5 = 3.9

11) 1.7 + ___ = 4.98

12) 5.25 + ___ = 7

13) 6.55 − ___ = 2.45

14) ___ − 3.98 = 5.32

15) ___ − 11.67 = 14.48

16) 12.35 + ___ = 14.78

17) ___ − 23.89 = 13.90

18) ___ + 17.28 = 19.56

19) 77.90 + ___ = 102.60

Multiplying and Dividing Decimals

✏ *Find the product.*

1) $0.5 \times 0.4 =$

2) $2.5 \times 0.2 =$

3) $1.25 \times 0.5 =$

4) $0.75 \times 0.2 =$

5) $1.92 \times 0.8 =$

6) $0.55 \times 0.4 =$

7) $3.24 \times 1.2 =$

8) $12.5 \times 4.2 =$

9) $22.6 \times 8.2 =$

10) $17.2 \times 4.5 =$

11) $25.1 \times 12.5 =$

12) $33.2 \times 2.2 =$

✏ *Find the quotient.*

13) $1.67 \div 100 =$

14) $52.2 \div 1,000 =$

15) $4.2 \div 2 =$

16) $8.6 \div 0.5 =$

17) $12.6 \div 0.2 =$

18) $16.5 \div 5 =$

19) $13.25 \div 100 =$

20) $25.6 \div 0.4 =$

21) $28.24 \div 0.1 =$

22) $34.16 \div 0.25 =$

23) $44.28 \div 0.5 =$

24) $38.78 \div 0.02 =$

Answers of Worksheets – Chapter 17

Decimal Place Value

1) ones
2) hundredths
3) hundredths
4) tens
5) tenths
6) thousands
7) hundredths
8) tenths
9) hundreds
10) ones
11) ones
12) tens
13) thousands
14) hundredths
15) tens
16) ones
17) tenths
18) hundreds
19) 0.01
20) 10
21) 0.2
22) 700
23) 30
24) 5
25) 0.05
26) 2,000
27) 400
28) 0.2

Order and Comparing Decimals

1) >
2) <
3) <
4) >
5) >
6) <
7) =
8) <
9) >
10) >
11) =
12) <
13) <
14) >
15) >
16) =
17) >
18) <
19) 0.23, 0.36, 0.4, 0.54, 0.87
20) 1.2, 1.80, 1.97, 2.4, 3.65
21) 0.34, 0.67, 1.2, 1.9, 2.3
22) 1.2, 1.34, 1.7, 3.2, 3.55, 4.2

Adding and Subtracting Decimals

1) 19.68
2) 59.72
3) 96.28
4) 12.28
5) 87.72
6) 15.38
7) 43.13
8) 52.60
9) 144.58
10) 1.4
11) 3.28
12) 1.75
13) 4.1
14) 9.3
15) 26.15
16) 2.43
17) 37.79
18) 2.28
19) 24.7

Multiplying and Dividing Decimals

1) 0.2
2) 0.5
3) 0.625
4) 0.15
5) 1.536
6) 0.22
7) 3.888
8) 52.5
9) 185.32

10) 77.4	15) 2.1	20) 64
11) 313.75	16) 4.3	21) 282.4
12) 73.04	17) 63	22) 136.64
13) 0.0167	18) 3.3	23) 88.56
14) 0.0522	19) 0.1325	24) 1,939

STAAR Test Review

The State of Texas Assessments of Academic Readiness (STAAR) is developed under the supervision of the Texas Education Agency and is taken by all public school students in Texas, grades 3–12. The tests measure the progress of students from 3rd grade to 8th grade, as well as high school. STAAR is the state's testing program and is based on state curriculum standards in core subjects including:

- o Reading,
- o Writing,
- o Mathematics,
- o Science,
- o Social Studies

In high school, students take end-of-course STAAR exams in five high school subjects:

- o Algebra I,
- o Biology,
- o English I,
- o English II,
- o U.S. History.

Students take STAAR tests in the spring. The number of tests a student takes each year will depend on what grade he or she is in. Most students will have two to four testing days during a school year.

In this book, we have reviewed all mathematics topics being covered on the STAAR test for grade 5. In this section, there are two complete Grade 5 STAAR Math Tests. Take these tests to see what score you'll be able to receive on a real STAAR Math test.

Good luck!

STAAR Mathematics Practice Tests

Time to Test

Time to refine your skill with a practice examination

Take a practice STAAR Math Test for grade 5 to simulate the test day experience. After you've finished, evaluate your test using the answer key.

Before You Start

- You'll need a pencil and scratch papers to take the test.

- For each multiple-choice question, there are four possible answers. Choose which one is best. For grids in questions, write your answer in the box provided.

- It's okay to guess. You won't lose any points if you're wrong.

- After you've finished the test, review the answer key to see where you went wrong.

- **Calculators are NOT allowed for the STAAR Test Grade 5.**

Good Luck

STAAR Practice

Test 1

State of Texas Assessments of

Academic Readiness

Grade 5

Mathematics

2019

1) The area of a circle is 81π. What is the circumference of the circle?

A. 9π

B. 18π

C. 36π

D. 81π

2) A rope weighs 600 grams per meter of length. What is the weight in kilograms of 11.3 meters of this rope? ($1\ kilograms = 1000\ grams$)

A. 0.0678

B. 0.678

C. 6.78

D. $6,780$

3) Solve.

$$\frac{1}{2} + \frac{4}{5} - \frac{3}{10} =$$

A. $\dfrac{9}{10}$

B. $\dfrac{2}{10}$

C. 1

D. 2

4) How many $\frac{1}{4}$ cup servings are in a package of cheese that contains $6\frac{1}{2}$ cups altogether?

Write your answer in the box below.

5) With what number must 8.421321 be multiplied in order to obtain the number 84,213.21?

A. 100

B. 1,000

C. 10,000

D. 100,000

6) Lily and Ella are in a pancake–eating contest. Lily can eat two pancakes per minute, while Ella can eat $2\frac{1}{2}$ pancakes per minute. How many total pancakes can they eat in 5 minutes?

A. 9.5 Pancakes

B. 29.5 Pancakes

C. 22.5 Pancakes

D. 11.5 Pancakes

7) The distance between cities A and B is approximately 2,600 miles. If Alice drive an average of 68 miles per hour, how many hours will it take Alice to drive from city A to city B?

A. Approximately 41 Hours

B. Approximately 38 Hours

C. Approximately 29 Hours

D. Approximately 27 Hours

8) 12 yards 4 feet and 2 inches equals to how many inches?

A. 96

B. 432

C. 482

D. 578

9) Which expression has a value of −8?

A. $8 − (−2) + (−18)$

B. $2 + (− 3) × (−2)$

C. $−6 × (−6) + (−2) × (−12)$

D. $(− 2) × (−7) + 4$

10) The drivers at G & G trucking must report the mileage on their trucks each week. The mileage reading of Ed's vehicle was 40,907 at the beginning of one week, and 41,053 at the end of the same week. What was the total number of miles driven by Ed that week?

A. 46 miles

B. 145 miles

C. 146 miles

D. 1,046 miles

11) In a bag, there are 72 cards. Of these cards, 12 cards are white. What fraction of the cards are white?

A. $\frac{1}{6}$

B. $\frac{2}{5}$

C. $\frac{12}{55}$

D. $\frac{1}{5}$

12) Which expression is equal to $\frac{3}{11}$?

A. $3 - 11$

B. $3 \div 11$

C. 3×11

D. $\frac{11}{3}$

13) If $A = 20$, then which of the following equations are correct?

A. $A + 20 = 40$

B. $A \div 20 = 40$

C. $20 \times A = 40$

D. $A - 20 = 40$

14) How long does a 270–miles trip take moving at 60 miles per hour (mph)?

A. 3 hours

B. 4 hours and 30 minutes

C. 6 hours and 20 minutes

D. 4 hours and 30 minutes

15) 50 is What percent of 40?

A. 20%

B. 25%

C. 125%

D. 150%

16) The perimeter of the trapezoid below is 52. What is its area?

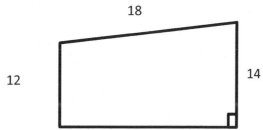

Write your answer in the box below.

---

17) Solve.

$$\frac{5}{8} \times \frac{4}{5} =$$

A. $\frac{1}{2}$

B. $\frac{10}{40}$

C. $\frac{20}{60}$

D. $\frac{1}{4}$

18) A cereal box has a height of 32 centimeters. The area of the base is 110 centimeters. What is the volume of the cereal box?

Write your answer in the box below.

---

19) Nancy ordered 18 pizzas. Each pizza has 8 slices. How many slices of pizza did Nancy ordered?

A. 124

B. 144

C. 156

D. 180

20) William keeps track of the length of each fish that he catches. Following are the lengths in inches of the fish that he caught one day: 13, 14, 9, 11, 9, 10, 18

What is the median fish length that William caught that day?

A. 11 *inches*

B. 9 *inches*

C. 12 *inches*

D. 13 *inches*

21) $9 + [8 \times 5] \div 2 = ?$

Write your answer in the box below.

22) What is the median of these numbers? 6, 3, 12, 8, 18, 21, 14

A. 12

B. 6

C. 18

D. 21

23) Camille uses a 30% off coupon when buying a sweater that costs $50. How much does she pay?

A. $35

B. $40

C. $42.50

D. $45

24) A baker uses 4 eggs to bake a cake. How many cakes will he be able to bake with 188 eggs?

A. 46

B. 47

C. 48

D. 49

25) Which of the following angles is obtuse?

A. 25 Degrees

B. 49 Degrees

C. 78 Degrees

D. 115 Degrees

26) Which of the following fractions is the largest?

A. $\dfrac{5}{8}$

B. $\dfrac{3}{7}$

C. $\dfrac{8}{9}$

D. $\dfrac{5}{11}$

27) The area of a rectangle is D square feet and its length is 9 feet. Which equation represents W, the width of the rectangle in feet?

A. $W = \dfrac{D}{9}$

B. $W = \dfrac{9}{D}$

C. $W = 9D$

D. $W = 9 + D$

28) Which list shows the fractions in order from least to greatest?

$$\frac{2}{3}, \frac{5}{7}, \frac{3}{10}, \frac{1}{2}, \frac{6}{13}$$

A. $\dfrac{2}{3}, \dfrac{5}{7}, \dfrac{3}{10}, \dfrac{1}{2}, \dfrac{6}{13}$

B. $\dfrac{6}{13}, \dfrac{1}{2}, \dfrac{2}{3}, \dfrac{5}{7}, \dfrac{3}{10}$

C. $\dfrac{3}{10}, \dfrac{2}{3}, \dfrac{5}{7}, \dfrac{1}{2}, \dfrac{6}{13}$

D. $\dfrac{3}{10}, \dfrac{6}{13}, \dfrac{1}{2}, \dfrac{2}{3}, \dfrac{5}{7}$

29) Which statement about 5 multiplied by $\frac{2}{3}$ is true?

A. The product is between 2 and 3

B. The product is between 3 and 4

C. The product is more than $\frac{11}{3}$

D. The product is between $\frac{14}{3}$ and 5

30) What is the volume of this box?

A. $30 \ cm^3$

B. $42 \ cm^3$

C. $35 \ cm^3$

D. $210 \ cm^3$

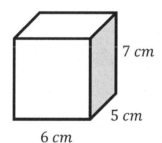

7 cm

5 cm

6 cm

31) A shirt costing $200 is discounted 15%. Which of the following expressions can be used to find the selling price of the shirt?

A. $(200)(0.70)$

B. $(200) - 200(0.30)$

C. $(200)(0.15) - (200)(0.15)$

D. $(200)(0.85)$

32) The length of a rectangle is $\frac{3}{4}$ of inches and the width of the rectangle is $\frac{5}{6}$ of inches. What is the area of that rectangle?

A. $\frac{1}{2}$

B. $\frac{5}{8}$

C. $\frac{20}{24}$

D. $\frac{5}{24}$

33) What is the volume of this box?

A. $24 \ cm^3$

B. $32 \ cm^3$

C. $162 \ cm^3$

D. $192 \ cm^3$

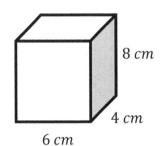

8 cm

4 cm

6 cm

34) How many square feet of tile is needed for a 18 feet to 18 feet room?

A. 72 Square Feet

B. 108 Square Feet

C. 216 Square Feet

D. 324 Square Feet

35) Of the 2,400 videos available for rent at a certain video store, 600 are comedies. What percent of the videos are comedies?

A. $18\frac{1}{2}\%$

B. 20%

C. 22%

D. 25%

36) How many 3 × 3 squares can fit inside a rectangle with a height of 54 and width of 12?

A. 72

B. 62

C. 50

D. 44

37) *ABC* Corporation earned only $200,000 during the previous year, two–third only of the management's predicted income. How much earning did the management predict?

A. $20,000

B. $30,000

C. $300,000

D. $340,000

38) The area of the base of the following cylinder is 50 square inches and its height is 10 inches.

What is the volume of the cylinder?

Write your answer in the box below.

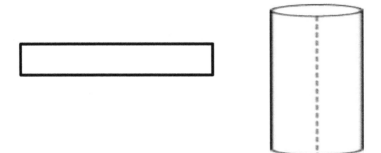

39) A rope 13 yards long is cut into 4 equal parts. Which expression does NOT equal to the length of each part?

A. $13 \div 4$

B. $\frac{13}{4}$

C. $4 \div 13$

D. $4\overline{)13}$

40) Calculate the area of the trapezoid in the following figure.

A. $4.5 \ ft^2$

B. $6.5 \ ft^2$

C. $13 \ ft^2$

D. $26 \ ft^2$

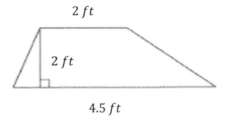

This is the End of Practice Test 1.   STOP

STAAR Practice Test 2

State of Texas Assessments of Academic Readiness

Grade 5

Mathematics

2019

1) If $x = -8$, which equation is true?

A. $x(2x - 4) = 120$

B. $8(4 - x) = 96$

C. $2(4x + 6) = 79$

D. $6x - 2 = -46$

2) A circle has a diameter of 8 inches. What is its approximate circumference?

$(\pi = 3.14)$

A. $6.28 \; inches$

B. $25.12 \; inches$

C. $34.85 \; inches$

D. $35.12 \; inches$

3) A woman owns a dog walking business. If 3 workers can walk 9 dogs, how many dogs can 5 workers walk?

A. 13

B. 15

C. 17

D. 19

4) What are the coordinates of the intersection of $x-axis$ and the $y-axis$ on a coordinate plane?

A. $(5,5)$
B. $(1,1)$
C. $(0,0)$
D. $(0,1)$

5) Jack added 19 to the product of 16 and 26. What is this sum?

A. 61
B. 330
C. 435
D. 7,904

6) Joe makes \$4.75 per hour at his work. If he works 8 hours, how much money will he earn?

A. \$32.00
B. \$34.75
C. \$36.50
D. \$38.00

7) Which of the following is an obtuse angle?

A. 89°

B. 55°

C. 143°

D. 235°

8) What is the value of $6 - 3\frac{4}{9}$?

A. $\dfrac{23}{9}$

B. $3\dfrac{4}{9}$

C. $-\dfrac{1}{9}$

D. $\dfrac{42}{9}$

9) The bride and groom invited 220 guests for their wedding. 190 guests arrived. What percent of the guest list was not present?

A. 90%

B. 20%

C. 23.32%

D. 13.64%

10) Frank wants to compare these two measurements.

$18.023\ kg$ ☐ $18,023\ g$

Which symbol should he use?

A. $<$

B. $>$

C. $\neq$

D. $=$

11) How long is the line segment shown on the number line below?

A. 6

B. 7

C. 8

D. 9

12) If a rectangle is 30 feet by 45 feet, what is its area?

A. 1,350

B. 1,250

C. 1,000

D. 870

13) If a vehicle is driven 32 miles on Monday, 35 miles on Tuesday, and 29 miles on Wednesday, what is the average number of miles driven each day?

A. 32 *miles*

B. 33 *miles*

C. 34 *miles*

D. 35 *miles*

14) Peter traveled 120 miles in 4 hours and Jason traveled 160 miles in 8 hours. What is the ratio of the average speed of Peter to average speed of Jason?

A. 3 : 2

B. 2 : 3

C. 5 : 9

D. 5 : 6

15) Aria was hired to teach three identical 5th grade math courses, which entailed being present in the classroom 36 hours altogether. At \$25 per class hour, how much did Aria earn for teaching one course?

A. \$50

B. \$300

C. \$600

D. \$1,400

16) In a classroom of 60 students, 22 are male. What percentage of the class is female?

A. 51%

B. 59%

C. 63%

D. 73%

17) You are asked to chart the temperature during an 8–hour period to give the average. These are your results:

7 am: 2 degrees

8 am: 5 degrees

9 am: 22 degrees

10 am: 28 degrees

11 am: 32 degrees

12 pm: 35 degrees

1 pm: 35 degrees

2 pm: 33 degrees

What is the average temperature?

A. 24

B. 28

C. 36

D. 46

18) What is 5,231.48245 rounded to the nearest tenth?

A. 5,231.482

B. 5,231.5

C. 5,231

D. 5,231.48

19) A car uses 15 gallons of gas to travel 450 miles. How many miles per gallon does the car get?

A. 26 miles per gallon

B. 28 miles per gallon

C. 30 miles per gallon

D. 34 miles per gallon

20) Number 0.025 is equal to what percent?

A. 0.03%

B. 0.25%

C. 2.50%

D. 25%

21) If one acre of forest contains 153 pine trees, how many pine trees are contained in 32 acres?

A. 4,896

B. 4,602

C. 4,308

D. 4,062

22) In a party, 6 soft drinks are required for every 9 guests. If there are 171 guests, how many soft drinks are required?

A. 9
B. 27
C. 114
D. 171

23) If Ella needed to buy 6 bottles of soda for a party in which 10 people attended, how many bottles of soda will she need to buy for a party in which 5 people are attending?

A. 3

B. 6

C. 9

D. 12

24) Julie gives 8 pieces of candy to each of her friends. If Julie gives all her candy away, which amount of candy could have been the amount she distributed?

A. 187

B. 216

C. 243

D. 223

25) Three co–workers contributed $10.25, $11.25, and $18.45 respectively to purchase a retirement gift for their boss. What is the maximum amount they can spend on a gift?

A. $20.25

B. $27.17

C. $39.95

D. $47.06

26) A rectangular plot of land is measured to be 160 feet by 200 feet. What is its total area?

A. 32,000 square feet

B. 28,404 square feet

C. 20,200 square feet

D. 2,040 square feet

27) A barista averages making 12 cups of coffee per hour. At this rate, how many hours will it take until she's made 960 cups of coffee?

A. 75

B. 80

C. 85

D. 90

28) Ava needs $\frac{1}{5}$ of an ounce of salt to make 1 cup of dip for fries. How many cups of dip will she be able to make if she has 50 ounces of salt?

A. 35

B. 55

C. 75

D. 250

29) While at work, Emma checks her email once every 90 minutes. In 9 hours, how many times does she check her email?

A. 4 Times

B. 5 Times

C. 6 Times

D. 7 Times

30) In a classroom of 44 students, 18 are male. About what percentage of the class is female?

A. 63%

B. 51%

C. 59%

D. 53%

31) A florist has 516 flowers. How many full bouquets of 12 flowers can he make?

A. 40

B. 41

C. 43

D. 45

32) Five out of 30 students had to go to summer school. What is the ratio of students who did not have to go to summer school expressed, in its lowest terms?

A. $\frac{5}{6}$

B. $\frac{7}{8}$

C. $\frac{3}{4}$

D. $\frac{6}{7}$

33) Which of the following fractions is the largest?

A. $\frac{2}{5}$

B. $\frac{1}{3}$

C. $\frac{6}{9}$

D. $\frac{5}{7}$

34) If 4 garbage trucks can collect the trash of 28 homes in a day. How many trucks are needed to collect in 70 houses?

A. 8
B. 9
C. 5
D. 10

35) A steak dinner at a restaurant costs $8.25. If a man buys a steak dinner for himself and 3 friends, what will the total cost be?

A. $27.00
B. $17.01
C. $33.00
D. $34.50

36) David's motorcycle stalled at the beach and he called the towing company. They charged him $3.45 per mile for the first 22 miles and then $4.25 per mile for each mile over 22. David was 26 miles from the motorcycle repair shop. How much was David's towing bill?

A. $71.40

B. $81.90

C. $90.90

D. $92.90

37) If a rectangular swimming pool has a perimeter of 112 feet and it is 22 feet wide, what is its area?

A. 1,896 square feet

B. 1,600 square feet

C. 1,464 square feet

D. 748 square feet

38) What is the volume of the following rectangle prism?

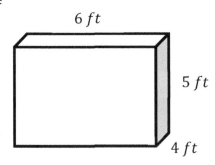

A. $15 \, ft^3$

B. $20 \, ft^3$

C. $24 \, ft^3$

D. $120 \, ft^3$

39) Which list shows the fractions listed in order from least to greatest?

$$\frac{1}{6} \qquad \frac{1}{8} \qquad \frac{1}{3} \qquad \frac{1}{10}$$

A.

$\frac{1}{3}$	$\frac{1}{6}$	$\frac{1}{8}$	$\frac{1}{10}$

B.

$\frac{1}{8}$	$\frac{1}{3}$	$\frac{1}{10}$	$\frac{1}{6}$

C.

$\frac{1}{6}$	$\frac{1}{10}$	$\frac{1}{3}$	$\frac{1}{8}$

D.

$\frac{1}{10}$	$\frac{1}{8}$	$\frac{1}{6}$	$\frac{1}{3}$

40) In a triangle ABC the measure of angle ACB is 25° and the measure of angle CAB is 45°. What is the measure of angle ABC?

Write your answer in the box below.

This is the End of Practice Test 2.   STOP

STAAR Practice Tests Answers and Explanations

STAAR Math Practice Test 1				STAAR Math Practice Test 2			
1	B	21	29	1	B	21	C
2	C	22	B	2	B	22	A
3	C	23	A	3	B	23	B
4	26	24	B	4	C	24	C
5	C	25	D	5	C	25	C
6	C	26	C	6	D	26	B
7	B	27	A	7	C	27	D
8	C	28	D	8	A	28	C
9	A	29	B	9	D	29	C
10	C	30	D	10	D	30	C
11	A	31	D	11	D	31	A
12	B	32	B	12	A	32	D
13	A	33	D	13	A	33	D
14	D	34	D	14	A	34	C
15	C	35	D	15	B	35	D
16	104	36	A	16	C	36	D
17	A	37	C	17	A	37	D
18	3,520	38	500	18	B	38	D
19	B	39	C	19	C	39	D
20	A	40	B	20	A	40	110

STAAR Practice Tests 1 Explanations

1) Choice B is correct.

Use area and circumference of circle formula.

$Area\ of\ a\ circle = \pi r^2 \Rightarrow 81\pi = \pi r^2 \Rightarrow r = 9$

$Circumference\ of\ a\ circle = 2\pi r \Rightarrow C = 2 \times 9 \times \pi \Rightarrow C = 18\pi$

2) Choice C is correct.

$1\ meter\ of\ the\ rope = 600\ grams$

$12.2\ meter\ of\ the\ rope = 11.3 \times 600 = 6,780\ grams = 6.78\ kilograms$

3) Choice C is correct.

$\frac{1}{2} + \frac{4}{5} - \frac{3}{10} = \frac{(5\times1)+(2\times4)-(1\times3)}{10} = \frac{10}{10} = 1$

4) Answer is 26.

To solve this problem, divide $6\frac{1}{2}$ by $\frac{1}{4}$. $\qquad 6\frac{1}{2} \div \frac{1}{4} = \frac{13}{2} \div \frac{1}{4} = \frac{13}{2} \times \frac{4}{1} = 26$

5) Choice C is correct.

The question is that number 84,213.21 is how many times of number 8.421321. The answer is 10,000.

6) Choice C is correct.

Lily eats 2 pancakes in 1 minute $\Rightarrow$ Lily eats 2×5 pancakes in 5 minutes.

Ella eats $2\frac{1}{2}$ pancakes in 1 minute $\Rightarrow$ Ella eats $2\frac{1}{2} \times 5$ pancakes in 5 minutes.

In total Lily and Ella eat $10 + 12.5$ pancakes in 5 minutes.

7) Choice B is correct.

Alice drives 68 miles in one hour. Therefore, she drives 2600 miles in about $(2600 \div 68)38$ hours.

8) **Choice C is correct.**

$12 \ yards = 12 \times 36 = 432 \ inches$

$4 \ feet = 4 \times 12 = 48 \ inches$

$12 \ yards \ 4 \ feet \ and \ 2 \ inches = 432 \ inches + 48 \ inches + 2 \ inches = 482 \ inches$

9) **Choice A is correct.**

Simplify each option provided using order of operations rules.

- A. $8 - (-2) + (-18) = 8 + 2 - 18 = -8$
- B. $2 + (-3) \times (-2) = 2 + 6 = 8$
- C. $-6 \times (-6) + (-2) \times (-12) = 36 + 24 = 60$
- D. $(-2) \times (-7) + 4 = 14 + 4 = 18$

Only option A is -8.

10) **Choice C is correct.**

To find the answer, subtract 40,907 from 41,053.

$41,053 - 40,907 = 146 \ miles$

11) **Choice A is correct.**

There are 65 cards in the bag and 13 of them are white. Then, 12 out of 72 cards are white. You can write this as: $\frac{12}{72}$. To simplify this fraction, divide both numerator and denominator by 13. Then:

$$\frac{12}{72} = \frac{1}{6}$$

12) **Choice B is correct.**

$\frac{3}{11}$ means 3 is divided by 11. The fraction line simply means division or $\div$. Therefore, we can write $\frac{3}{11}$ as $3 \div 11$.

13) **Choice A is correct.**

Plug in 20 for A in the equations. Only option A works.

$A + 20 = 40$

$20 + 20 = 40$

14)   **Choice D is correct.**

$60\ miles : 1\ hour$

$270\ miles : 270 \div 60 = 4.5\ hours$

15)   **Choice C is correct.**

Use percent formula:

$\text{part} = \frac{\text{percent}}{100} \times \text{whole}$

$50 = \frac{\text{percent}}{100} \times 40 \Rightarrow 50 = \frac{\text{percent} \times 40}{100} \Rightarrow 50 = \frac{\text{percent} \times 4}{10}$, multiply both sides by 10.

$500 = percent \times 4$, divide both sides by 4.

$125 = percent$

16)   **Answer is 104.**

First, find the missing side of the trapezoid. The perimeter of the trapezoid below is 52.
Therefore, the missing side of the trapezoid (its height) is: $52 - 12 - 18 - 14 = 52 - 44 = 8$

Area of a trapezoid: $A = \frac{1}{2}h(b_1 + b_2) = \frac{1}{2}(8)(12 + 14) = 104$

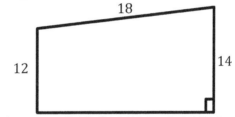

17) Choice A is correct.

$\frac{5}{8} \times \frac{4}{5} = \frac{5 \times 4}{8 \times 5} = \frac{20}{40} = \frac{1}{2}$

18) Answer is 3,520.

Use volume of cube formula.

$Volume = base \times height \Rightarrow V = 110 \times 32 \Rightarrow V = 3,520$

19) Choice B is correct.

1 pizza has 8 slices.

18 pizzas contain (18 × 8) 144 slices.

20) Choice A is correct.

Write the numbers in order:

$9, 9, 10, 11, 13, 14, 18$

Median is the number in the middle. Therefore, the median is 11.

21) Answer is 29.

Use PEMDAS (order of operation):

$9 + [8 \times 5] \div 2 = 9 + (40) \div 2 = 9 + (40 \div 2) = 29$

22) Choice B is correct.

Write the numbers in order: $3, 6, 8. 12, 14, 18, 21$

Median is the number in the middle. Therefore, the median is 12.

23) Choice A is correct.

Let x be the new price after discount.

$x = 50 \times (100 - 30)\% = 50 \times 70\% = 50 \times 0.70 = 35$

$x = \$35$

24) Choice B is correct.

4 eggs for 1 cake. Therefore, 188 eggs can be used for $(188 \div 4) 47$ cakes.

25) Choice D is correct.

An obtuse angle is any angle larger than 90 degrees. From the options provided, only D (115 degrees) is larger than 90.

26) Choice C is correct.

Compare the fractions.

$\frac{5}{8} > \frac{3}{7}$

and $\frac{8}{9} > \frac{5}{11}$

$\dfrac{8}{9} > \dfrac{5}{8}$

Therefore, $\dfrac{8}{9}$ is the biggest fraction.

27) Choice A is correct.

Use area of rectangle formula.

$$area\ of\ a\ rectangle\ =\ width\ \times\ length \Rightarrow D = w \times l \Rightarrow w = \dfrac{D}{l} = \dfrac{D}{9}$$

28) Choice D is correct.

To list the fractions from least to greatest, you can convert the fractions to decimal.

$$\dfrac{2}{3} = 0.67$$
$$\dfrac{5}{7} = 0.71$$
$$\dfrac{3}{10} = 0.3$$
$$\dfrac{1}{2} = 0.5$$
$$\dfrac{6}{13} = 0.46$$

$$\dfrac{3}{10} = 0.3, \dfrac{6}{13} = 0.46, \dfrac{1}{2} = 0.5, \dfrac{2}{3} = 0.67, \dfrac{5}{7} = 0.71$$

Option D shows the fractions in order from least to greatest.

29) Choice B is correct.

5 multiplied by $\dfrac{2}{3} = \dfrac{10}{3} = 3.33$, therefore, only choice B is correct.

30) Choice D is correct.

Use volume of rectangle formula.

$$Volume\ of\ a\ rectangle = width \times length \times heigh \Rightarrow V = 5 \times 6 \times 7 \Rightarrow V = 210$$

31) Choice D is correct.

To find the selling price, multiply the price by (100% − rate of discount).

Then: $(200)(100\% - 15\%) = (200)(0.85) = 170$

32) Choice B is correct.

Use area of rectangle formula.

$$Area = \ length \ \times \ width \ \Rightarrow A = \frac{3}{4} \times \frac{5}{6} \Rightarrow A = \frac{5}{8} \ inches$$

33) Choice D is correct.

Use volume of cube formula.

$$Voluem = length \times width \times height \ \Rightarrow V = 6 \times 4 \times 8 \Rightarrow V = 192 \ cm^3$$

34) Choice D is correct.

Find the area of the room which is a square. Use area of square formula.

$$S = a^2 \Rightarrow S = \ 18 \ feet \times 18 \ feet \ = \ 324 \ square \ feet$$

35) Choice D is correct.

Use percent formula:

$$part = \frac{percent}{100} \times whole$$
$$60 = \frac{percent}{100} \times 2400 \ \Rightarrow \ 600 = percent \times 24 \ \Rightarrow percent = 25$$

36) Choice A is correct.

Use area of rectangle formula.

$$A = a \times b \Rightarrow A = \ 54 \times 12 \ \Rightarrow A = \ 648$$

Divide the area by 9 ($3 \times 3 = 9$ squares) to find the number of squares needed. $648 \div 9 = 72$

37) Choice C is correct.

ABC Corporation's income $= \frac{2}{3}$ management's predicted income.

$\$200,000 = \frac{2}{3}$ management's predicted income

management's predicted income $= \$200,000 \times \frac{3}{2} = \$300,000$

38) **Answer is 500.**

Use volume of cylinder formula.

$Voluem = base \times heigth \Rightarrow V = 50 \times 10 \Rightarrow V = 500$

39) **Choice C is correct.**

13 yards long rope is cut into 4 equal parts. Therefore, 13 should be divided by 4.
Only option C is NOT 13 divided by 4. (It is 4 divided by 13)

40) **Choice B is correct.**

Use area of trapezoid formula.

Area of trapezoid $= \frac{1}{2} \times heigth \times (base\ 1 + base\ 2) \Rightarrow \frac{1}{2} \times 2 \times (2 + 4.5) = 6.5$

STAAR Practice Tests 2 Explanations

1) Choice B is correct

Plug in $x = -8$ in each equation.

 A. $x(2x - 4) = 120 \rightarrow (-8)(2(-8) - 4) = (-8) \times (-16 - 4) = 160$
 B. $8(4 - x) = 96 \rightarrow 8(4 - (-8)) = 8(12) = 96$
 C. $2(4x + 6) = 79 \rightarrow 2(4(-8) + 6) = 2(-32 + 6) = -52$
 D. $6x - 2 = -46 \rightarrow 6(-8) - 2 = -48 - 2 = -50$

Only option B is correct.

2) Choice B is correct

The diameter of the circle is 8 inches. Therefore, the radius of the circle is 4 inches.

Use circumference of circle formula. $C = 2\pi r \Rightarrow C = 2 \times 3.14 \times 4 \Rightarrow C = 25.12$

3) Choice B is correct.

3 workers can walk 9 dogs $\Rightarrow$ 1 workers can walk 3 dogs. 5 workers can walk (5×3) 15 dogs.

4) Choice C is correct

The horizontal axis in the coordinate plane is called the $x - axis$. The vertical axis is called the $y - axis$. The point at which the two axes intersect is called the origin. The origin is at 0 on the $x - axis$ and 0 on the $y - axis$.

5) Choice C is correct

$19 + (16 \times 26) = 19 + 416 = 435$

6) Choice D is correct

$1\ hour$: $4.75   8\ hours$: $8 \times \$4.75 = \$38$

7) Choice C is correct

An obtuse angle is an angle of greater than 90° and less than 180°. From the options provided, only option C (143 degrees) is an obtuse angle.

8) Choice A is correct

$$6 - 3\frac{4}{9} = \frac{54}{9} - \frac{31}{9} = \frac{23}{9}$$

9) Choice D is correct.

The number of guests that are not present are $(220 - 190)30$ out of $220 = \frac{30}{220}$

Change the fraction to percent:

$\frac{30}{220} \times 100\% = 13.64\%$

10) Choice D is correct.

Each kilogram is 1,000 grams.

18,023 grams $= \frac{18,023}{1,000} = 18.023$ kilograms. Therefore, two amounts provided are equal.

11) Choice D is correct.

The line segment is from 1 to -8. Therefore, the line is 9 units.

$1 - (-8) = 1 + 8 = 9$

12) Choice A is correct.

Use area of rectangle formula.

$Area = length \times width \Rightarrow A = 30 \times 45 \Rightarrow A = 1,350$

13) Choice A is correct.

$\text{average (mean)} = \frac{\text{sum of terms}}{\text{number of terms}} \Rightarrow \text{average} = \frac{32+35+29}{3} \Rightarrow average = 32$

14) Choice A is correct

Peter's speed $= \frac{120}{4} = 30$

Jason's speed $= \frac{160}{8} = 20$

$\frac{The\ average\ speed\ of\ peter}{The\ average\ speed\ of\ Jason} = \frac{30}{20}$ equals to: $\frac{3}{2}$ or $3 : 2$

15) Choice B is correct.

Aria teaches 36 hours for three identical courses. Therefore, she teaches 12 hours for each course. Aria earns \$25 per hour. Therefore, she earned \$300 (12×25) for each course.

16) Choice C is correct.

The number of female students in the class is $(60 - 22)$ 38 out of $60 = \frac{38}{60}$

Change the fraction to percent:

$\frac{38}{60} \times 100\% = 63\%$

17) Choice A is correct.

average (mean) $= \frac{\text{sum of terms}}{\text{number of terms}} \Rightarrow$ average $= \frac{2+5+22+28+32+35+35+33}{8} \Rightarrow average =$ 24

18) Choice B is correct.

Rounding decimals is similar to rounding other numbers. If the hundredths and thousandths places of a decimal is forty-nine or less, they are dropped, and the tenths place does not change. For example, rounding 0.843 to the nearest tenth would give 0.8. Therefore, 5,231.48245 rounded to the nearest tenth is 5,231.5 .

19) Choice C is correct

Write a proportion and solve.

15 $gallons$: 450 $miles$

1 $gallon$: $450 \div 15 = 30$ $miles$

20) Choice C is correct

$0.025 = \frac{25}{1000} = \frac{2.5}{100} = 2.5\%$

21) Choice A is correct

1 $acre$: 153 $pine\ trees$

32 $acres$: $153 \times 32 = 4,896$ $pine\ trees$

22) Choice C is correct.

Write a proportion and solve.

$\frac{6\ \text{soft drinks}}{9\ \text{guests}} = \frac{x}{171\ \text{guests}}$

$x = \frac{171 \times 6}{9} \Rightarrow x = 114$

23) Choice A is correct

Write a proportion and solve.

$$\frac{6 \text{ soft drinks}}{10 \text{ guests}} = \frac{x}{5 \text{ guests}}$$

$$x = \frac{5 \times 6}{10} \Rightarrow x = 3$$

24) Choice B is correct.

The number of candies should be a whole number! Check each option provided.

$A. \ 167 \div 8 = 23.375$

$B. \ 216 \div 8 = 27$

$C. \ 243 \div 8 = 30.375$

$D. \ 223 \div 8 = 27.875$

Only option B gives a whole number.

25) Choice C is correct.

They contributed: $\$10.25 + \$11.25 + \$18.45 = \$39.95$ in total, so the maximum amount that they can spend is the sum of their contribution.

26) Choice C is correct.

Use area of rectangle formula.

$$Area = \ length \times width \Rightarrow A = 160 \times 200 \Rightarrow A = 32,000$$

27) Choice B is correct.

$12 \ cups: 1 \ hour$

$960 \ cups: 960 \div 12 = 80 \ hours$

28) Choice D is correct.

Write a proportion and solve.

$$\frac{\frac{1}{5}}{50} = \frac{1}{x} \ \Rightarrow x \ = 50 \times 5 = 250$$

29) Choice C is correct.

Every 90 minutes Emma checks her email.

In 9 hours (540 minutes), Emma checks her email $(540 \div 90)6$ times.

30) Choice C is correct.

There are 44 students in the class. 18 of the are male and 26 of them are female.

26 out of 44 are female. Then:

$$\frac{26}{44} = \frac{x}{100} \rightarrow 2{,}600 = 44x \rightarrow x = 2{,}600 \div 44 \approx 59\%$$

31) Choice C is correct.

Divide the number flowers by 12 : $516 \div 12 = 43$

32) Choice A is correct

The students that had to go to summer school is 5 out of $30 = \frac{5}{30} = \frac{1}{6}$

Therefore $\frac{5}{6}$ students did not have to go to summer school.

33) Choice D is correct

$\frac{2}{5} = 0.4$

$\frac{1}{3} = 0.33$

$\frac{6}{9} = 0.67$

$\frac{5}{7} = 0.71$

$\frac{5}{7}$ is the largest.

34) Choice D is correct

4 garbage trucks can collect the trash of 28 homes. Then, one garbage truck can collect the trash of 7 homes. To collect trash of 70 houses, $10(70 \div 7)$ garbage trucks are required.

35) Choice C is correct

4 steak dinners $= 4 \times \$8.25 = \$33$

36) Choice D is correct

$3.45 per mile for the first 22 miles. Therefore, the cost for the first 22 miles is:

$22 \times \$3.45 = \$75.9$

$4.25 per mile for each mile over 22, therefore, 4 miles over 22 miles cost: $4 \times \$4.25 = \$17$

In total, David pays: $\$75.9 + \$17 = \$92.90$

37) Choice D is correct.

Perimeter of rectangle formula:

$P = 2 \, (length \; + \; width) \Rightarrow 112 = 2 \, (l + 22) \; \Rightarrow l = 34$

Area of rectangle formula: $A = length \times width \; \Rightarrow \; A = 34 \times 22 \; \Rightarrow \; A = 748$

38) Choice D is correct.

Use volume of rectangle prism formula.

$V = length \times width \times height \; \Rightarrow \; V = 6 \times 4 \; \times 5 \; \Rightarrow \; V = 120$

39) Choice D is correct

In fractions, when denominators increase, the value of fractions decrease and as much as numerators increase, the value of fractions increase. Therefore, the least one of this list is: $\frac{1}{10}$ and the greatest one of this list is: $\frac{1}{3}$

40) Answer is 110

All angles in every triangle add up to 180°. Let x be the angle ABC. Then:

$180 = 25 + 45 + x \Rightarrow x = 110°$

www.EffortlessMath.com

... So Much More Online!

✓ FREE Math lessons

✓ More Math learning books!

✓ Mathematics Worksheets

✓ Online Math Tutors

Need a PDF version of this book?

Please visit www.EffortlessMath.com

Made in the USA
Coppell, TX
16 November 2020

41474237R00105